# ALL THUMBS Guide to Home Wiring

## Other All Thumbs Guides

Home Plumbing
Painting, Wallpapering and Stenciling
Repairing Major Home Appliances

# ALL THUMBS Guide to Home Wiring

Robert W. Wood
Illustrations by Steve Hoeft

## Notices

**National Electrical Code® (NEC)** National Fire Protection Assoc.
**Romex®** General Cable Co.

## Disclaimer

It is the intention of the author that the information presented is accurate and stresses the need for safety, however, due to the nature of electricity, neither the author nor TAB Books can be held liable with respect to the use of the information herein.

FIRST EDITION
SECOND PRINTING



**Library of Congress Cataloging-in-Publication Data**

Wood, Robert W., 1933-
Home wiring / by Robert W. Wood.
p. cm.
Includes index.
ISBN 0-8306-2544-5 (h) ISBN 0-8306-2543-7 (p)
1. Electric wiring, Interior—Amateurs' manuals. I. Title.
TK9901.W65 1992
621.319'24—dc20 91-33465
CIP

Acquisitions Editor: Kimberly Tabor
Designer: Jaclyn J. Boone
Editorial Team: Susan D. Wahlman, Editor
Joanne Slike
Production Team: Katherine G. Brown
Janice Ridenour, Layout
Jana L. Fisher, Typesetting
Brian Allison, Color Separation
Cover Design: Lori E. Schlosser
Cover Illustration: Denny Bond, East Petersburg, PA
Cartoon Caricature: Michael Malle, Pittsburgh, PA ATS

## The All Thumbs Guarantee

TAB Books/McGraw-Hill guarantees that you will be able to follow every step of each project in this book, from beginning to end, or you will receive your money back. If you are unable to follow the All Thumbs steps, return this book, your store receipt, and a brief explanation to:

All Thumbs
P.O. Box 581
Blue Ridge Summit, PA 17214-9998

## About the Binding

This and every All Thumbs book has a special lay-flat binding. To take full advantage of this binding, open the book to any page and run your finger along the spine, pressing down as you do so; the book will stay open at the page you've selected.

The lay-flat binding is designed to withstand constant use. Unlike regular book bindings, the spine will not weaken or crack when you press down on the spine to keep the book open.

# Contents

# Preface

A collection of books about do-it-yourself home repair and improvement, the All Thumbs series was created not for the skilled jack-of-all-trades, but for the average homeowner. If your familiarity with the various systems in the home is minimal, or your budget doesn't keep pace with today's climbing costs, this series is tailor-made for you.

Several different types of professional contractors are required to construct even the smallest home. Carpenters build the framework, plumbers install the pipes, and electricians complete the wiring. Few people can do it all. The necessary skills often require years to master. The professional works quickly and efficiently and depends on a large volume of work to survive. Because service calls are time-consuming, often requiring more travel time than actual labor, they can be expensive. The All Thumbs series saves you time and money by showing you how to make most common repairs yourself.

The guides cover topics such as home wiring; plumbing; painting, stenciling, and wallpapering; and repairing major appliances, to name a few. Copiously illustrated, each book details the procedures in an easy-to-follow, step-by-step format, making many repairs and home improvements well within the ability of nearly any homeowner.

# Introduction

Designed for the novice do-it-yourselfer, this book should eliminate the mystery and fear often associated with electricity. The step-by-step instructions and illustrations will show you how easy fixing some common electrical problems can be.

In chapter 1, I explain the home electrical system. In chapter 2, I describe the basic tools required and present a systematic approach to troubleshooting. This sequence is designed to guide you quickly to the source of each problem. Chapters 3 through 15 explain various repairs, with safe, step-by-step instructions and illustrations that will allow practically anyone to make the repairs.

The average homeowner's ability to perform electrical work is, of course, limited, but major electrical systems seldom break down. The most common problems usually can be handled by a beginning do-it-yourselfer. I'll start with a few very important do's and don'ts:

- Do turn off the power at the service panel before making repairs.
- Do work carefully.
- Do develop safe work habits.

- Don't work on a live circuit.
- Don't be careless.

Once you understand and respect the potential hazards, electrical wiring work is quite safe to do.

Electricity is a natural phenomenon that is everywhere. The most spectacular example is found in the atmosphere as lightning. Our earliest ancestors no doubt ran from their primitive camp sites to caves at the sudden thunderous bolts that lit up the sky. They probably cowered and wondered at this mystical force. Even today electricity can be terrifying to the uninformed. But if you have a little understanding of the basics, electricity can be your faithful servant.

Around the turn of the century, a code of safety standards was developed that became the National Electric Code. The NEC is now usually referred to as simply "the Code." It has been sponsored by the National Fire Protection Association since 1911. The Code is a set of rules, developed over the years, that have been found to provide safe and practical electrical installations. The Code does not have the force of law, but most state and local authorities do adopt it and might even include additional rules for their particular areas. The Code is updated and revised every three years. Copies are available at bookstores and libraries.

A safely installed wiring system can be defeated if any device or component is used that is manufactured below minimum safety standards. These standards were established by a nonprofit testing laboratory known as Underwriters Laboratories, Inc. (UL) or by the American National Standards Institute (ANSI).

Manufacturers submit their product to the laboratory for testing. If the product meets minimum safety standards, the product is listed by the Underwriters Laboratories and displays a UL label. The label means that, if it is used for the purpose for which it was intended, the device has passed the minimum safety requirements.

The UL symbol is not a reliable indicator of quality. The same type of product, made by a number of manufacturers, might each have the UL label, but one product of a higher quality might outlast a lower-quality product many times. Most stores sell only merchandise that is UL-listed, but you still need to shop for quality. When in doubt, seek professional advice.

CHAPTER ONE

# Understanding the Basics

Electric *current* moving along a wire is measured in units called *amperes*, usually abbreviated amps. In the United States, we use *alternating current* at a standard frequency of *60 cycles* per second. The force that moves the current is measured in *volts*. The work performed by the *voltage* and current is measured in *watts*. The most important condition that affects the flow of electrical current is *resistance*. Electrical resistance is similar to friction—the larger the resistance, the smaller the current flow. This relationship means that large wires must be used to transmit large currents. Resistance also tends to generate heat.

Electric companies generate electricity in huge plants, often in remote locations. The utility company delivers power to the residence through overhead wires or an underground service. The standard service has three wires: two *hot wires*, or *conductors*, and one *neutral wire*. Each hot wire supplies 120 volts when measured from the neutral wire. The neutral wire is kept at 0 volts and is considered to be at ground potential, or *ground*. With this arrangement, one hot wire and the neutral wire provide 120 volts for circuits used for lights and wall *outlets*. Both hot wires and the neutral provide 240 volts for large appliances such as air conditioners, ranges, and dryers.

*Electricity might travel many miles before it reaches your home.*

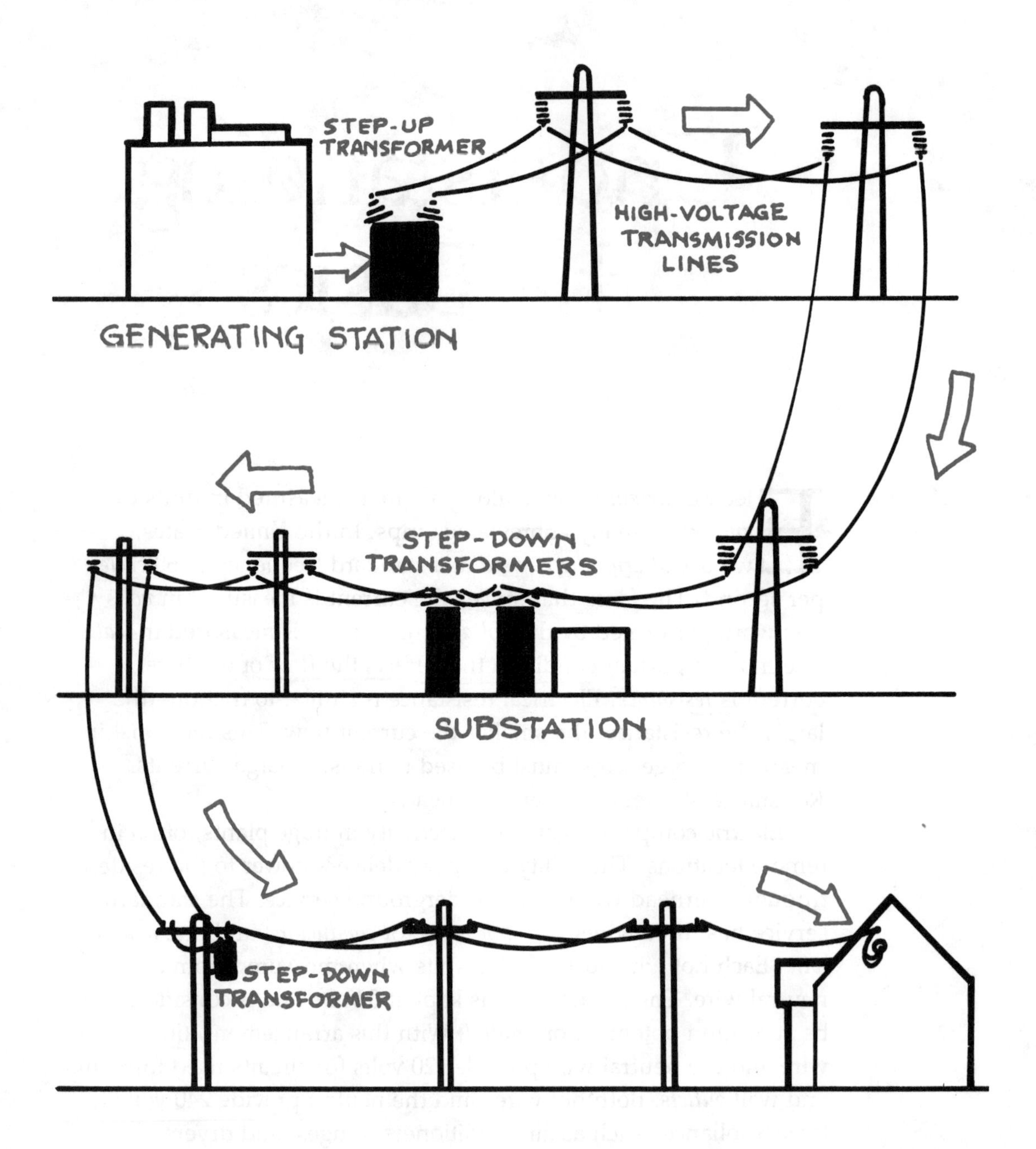

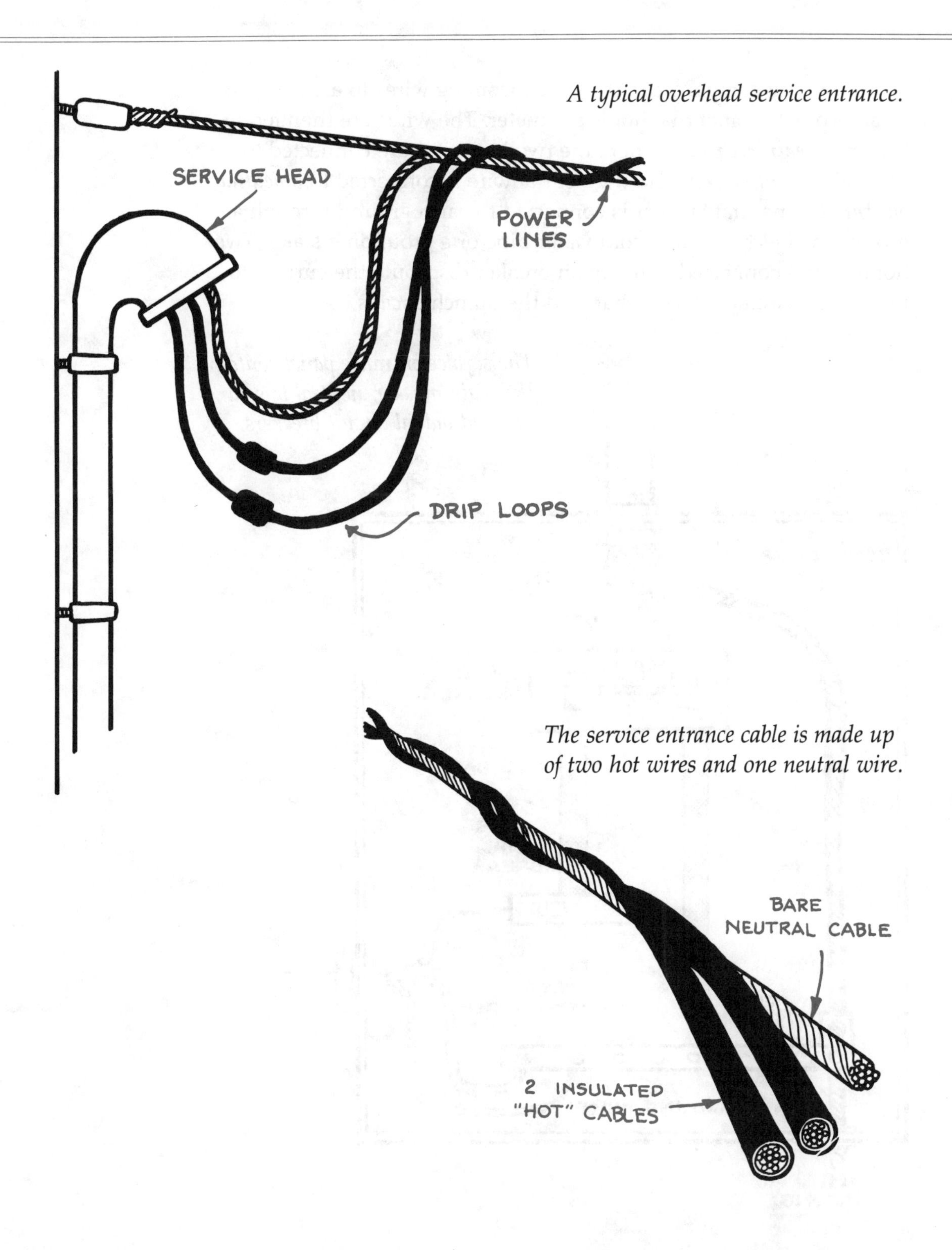


*A typical overhead service entrance.*

*The service entrance cable is made up of two hot wires and one neutral wire.*

The utility company connects the incoming wires to a weatherproof cabinet that holds the meter. The wires are then fed to the *service entrance panel*, where the two hot wires are connected to the main *circuit breaker*, and the neutral wire is connected to a *neutral bus bar*. The neutral bus bar is connected to earth ground through a grounded object such as a cold-water pipe or a grounding stake. Two hot bus bars connected to the main breaker distribute the current to the smaller circuit breakers that feed the branch *circuits*.

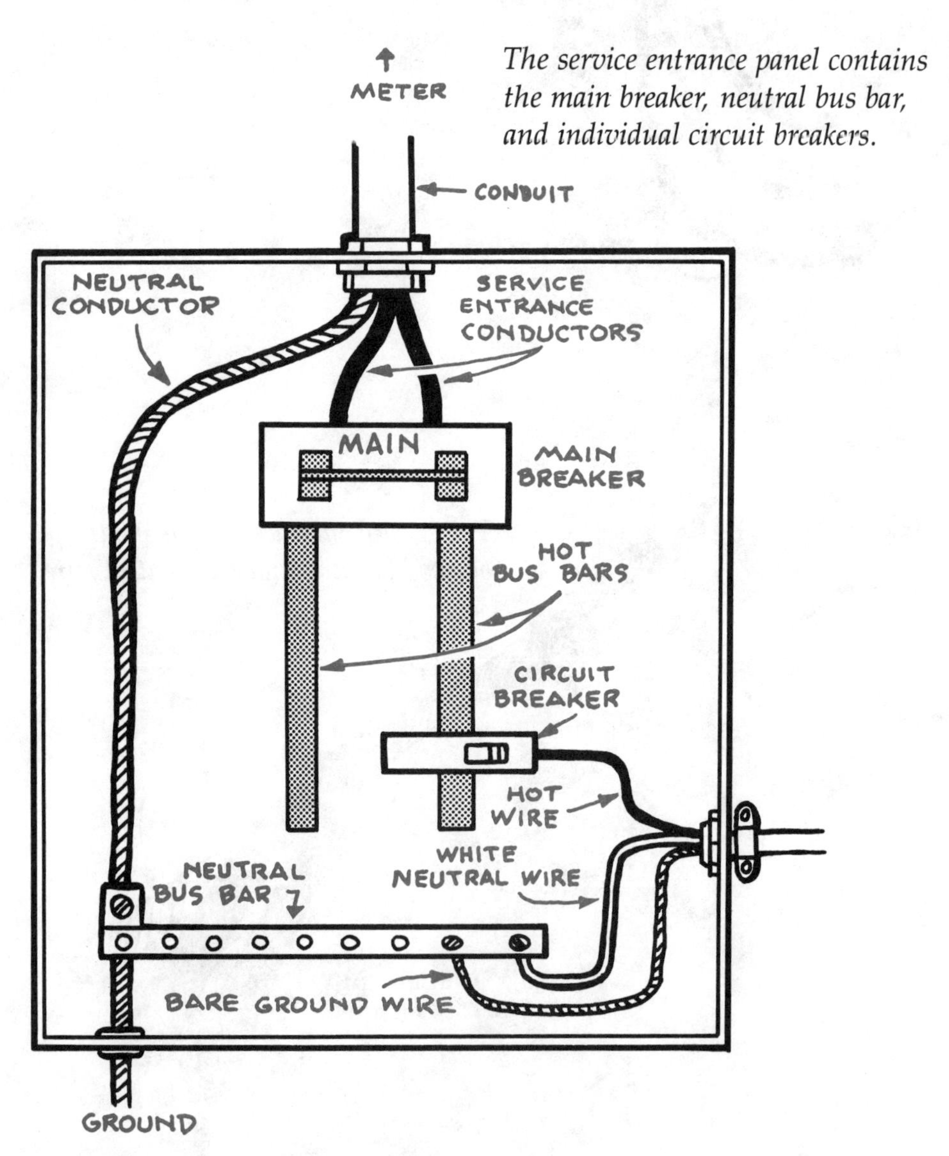


*The service entrance panel contains the main breaker, neutral bus bar, and individual circuit breakers.*

Several branch circuits might be needed to supply power to the kitchen. Part of the *cable* for the branch circuits is hidden in the attic. It might run through or across the ceiling joists. Other cables run through the studs in the walls. A circuit supplying power to lights could run through the attic to the light, then drop down for the *switch*.

*Kitchens require a number of circuits.*

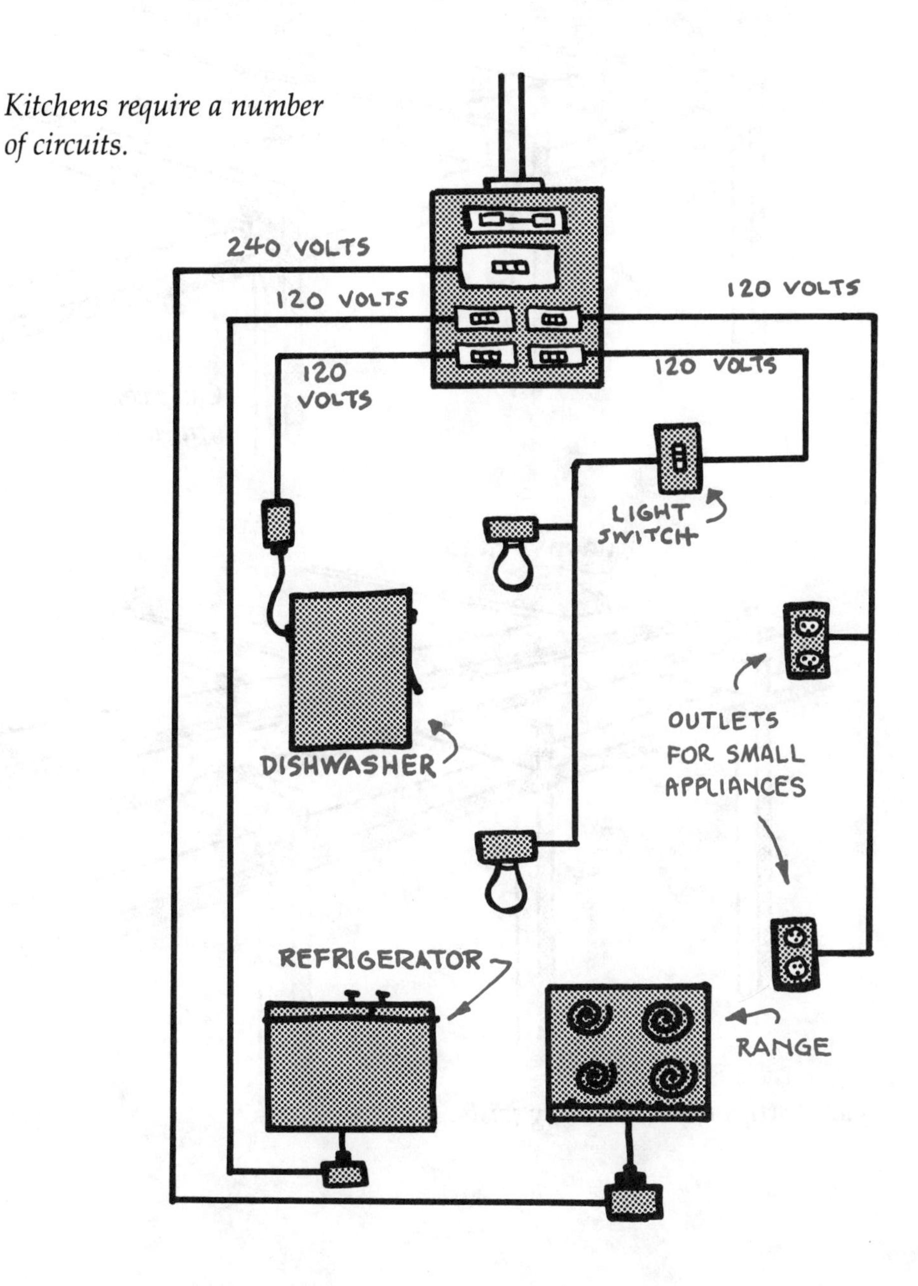

The cable for the *receptacles* should be fed by a different circuit. It also might run through the attic, then drop down into the walls to feed several receptacles.

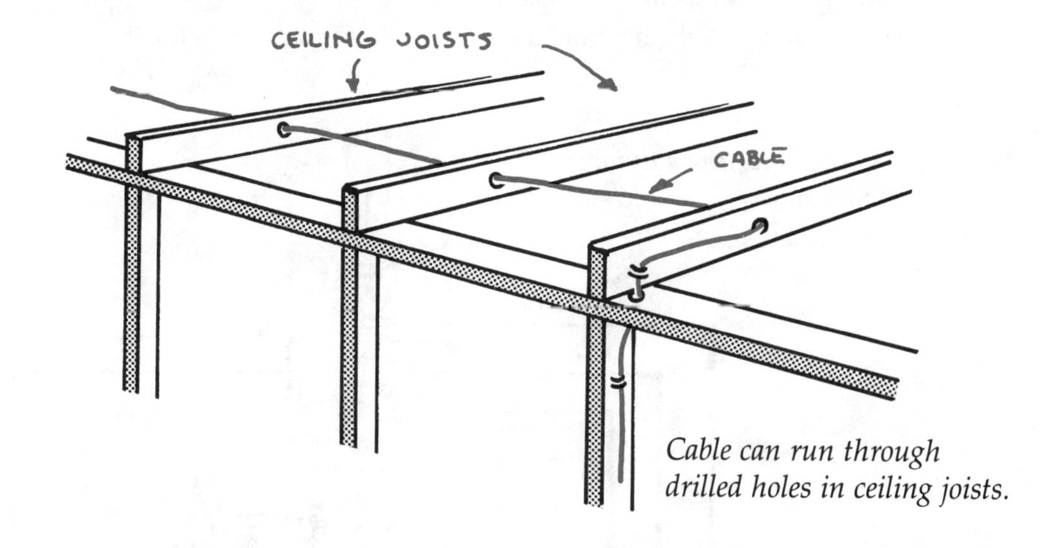


*Cable can run through drilled holes in ceiling joists.*

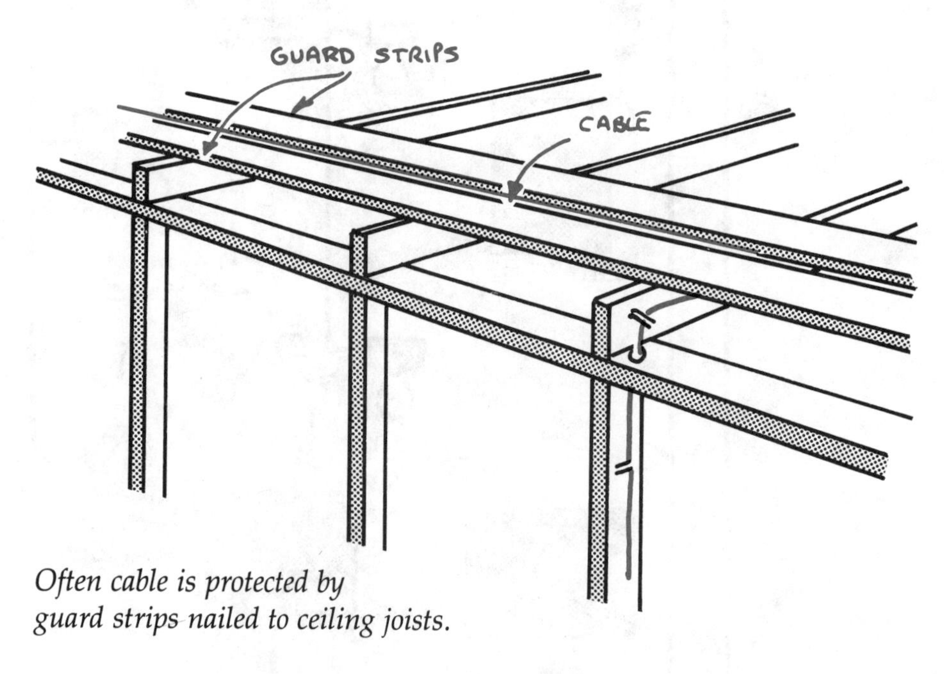


*Often cable is protected by guard strips nailed to ceiling joists.*

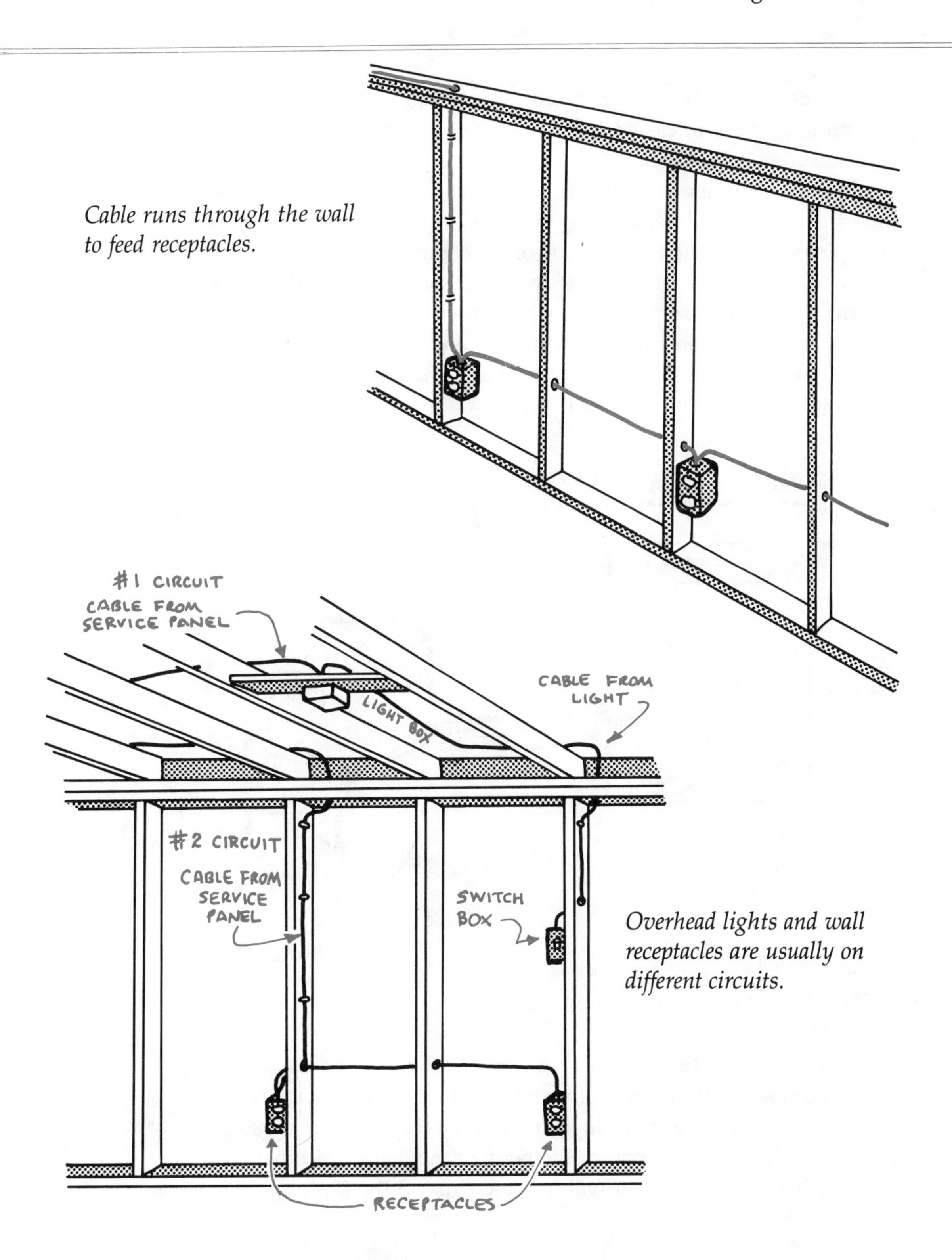


*Cable runs through the wall to feed receptacles.*

*Overhead lights and wall receptacles are usually on different circuits.*

In order for electricity to be used, the current must flow through a continuous path, or closed loop. This closed loop makes an electrical circuit. A branch circuit in the home begins at the smaller circuit breakers in the service entrance panel. Here, a hot wire runs from the breaker and then through the walls to an outlet, where, for example, it is connected to one side of a receptacle.

A neutral wire is connected to the other side of the receptacle and returns to the service entrance panel, where it is connected to the neutral bus bar. When a device such as a lamp is plugged into the outlet and the switch is turned on, the circuit is complete and current will flow. The light bulb will light.

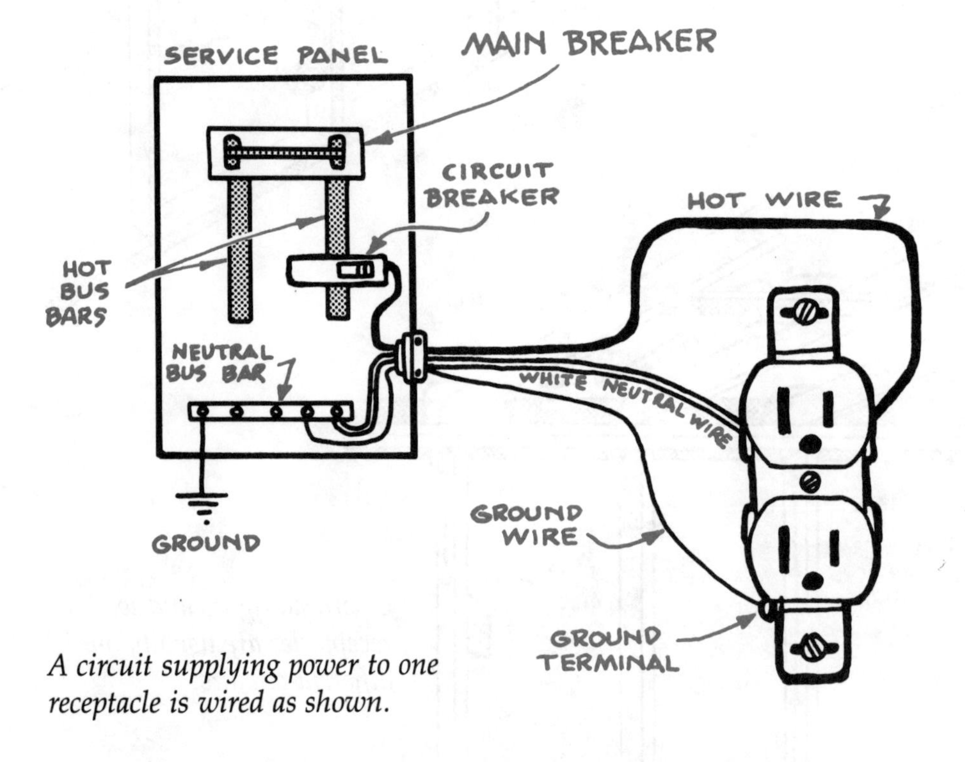


*A circuit supplying power to one receptacle is wired as shown.*

One other wire to consider is the ground wire. This wire, usually bare, is connected from the neutral bus bar to the metal housing of the receptacle. If the device plugged into the receptacle has a grounded plug, and any part of the body of the device becomes electrically alive, the breaker will trip and turn off the power.

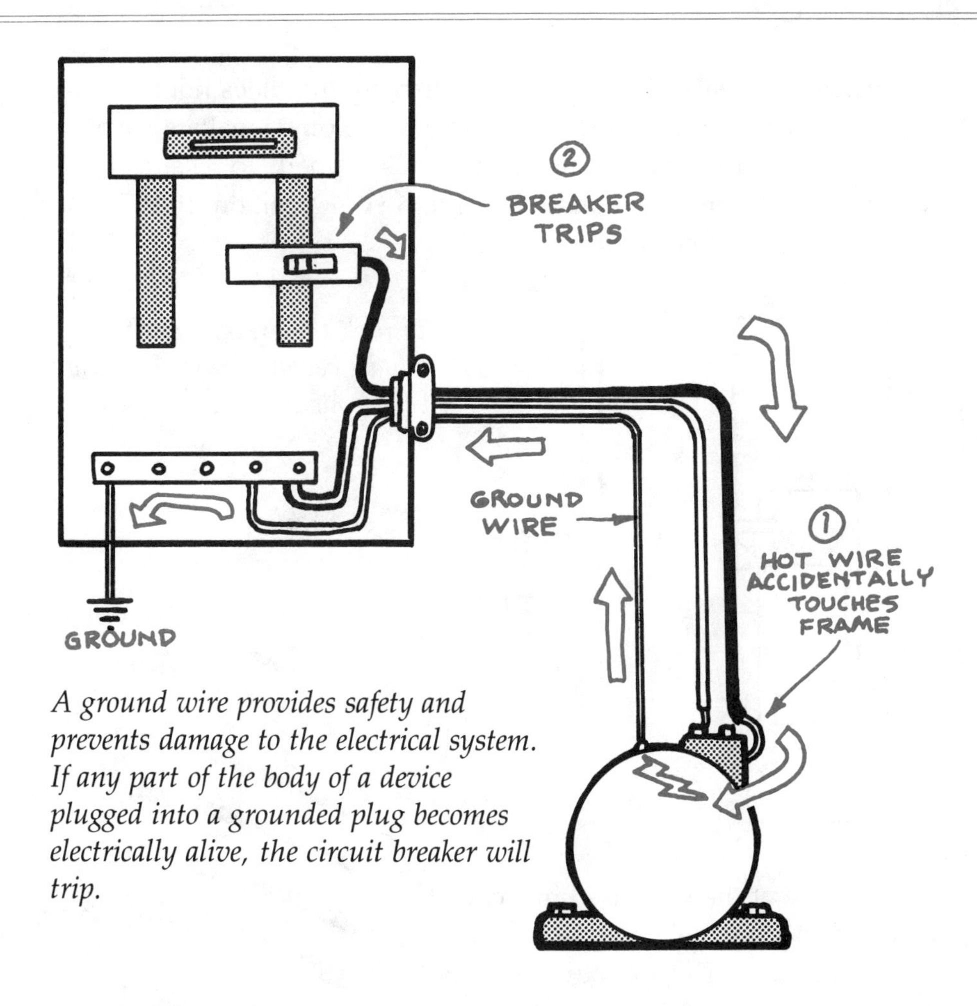


*A ground wire provides safety and prevents damage to the electrical system. If any part of the body of a device plugged into a grounded plug becomes electrically alive, the circuit breaker will trip.*

The purpose of a circuit breaker is to protect an electrical circuit from damage caused by too much current. By design, a breaker will trip if the circuit it is protecting is forced to carry more current than the wire can handle. These excessive currents are usually caused by a faulty appliance, a bad switch, or a bad receptacle. Also, too many appliances on one circuit will cause a breaker to trip.

Circuit breakers are rated for the same current as the wiring it is protecting. For example, #12 copper wire can safely handle 20 amps. The breaker protecting the circuit using #12 wire must never be larger than 20 amps. If it is, and a problem occurs, the wires could overheat and start a fire inside a wall or attic.

Inside the breaker is a tripping mechanism that does not let the breaker be reset until cooling has taken place. Some breakers have small windows that indicate the breaker has tripped. To reset a typical breaker, push the handle beyond the OFF position, then return it to the ON position.

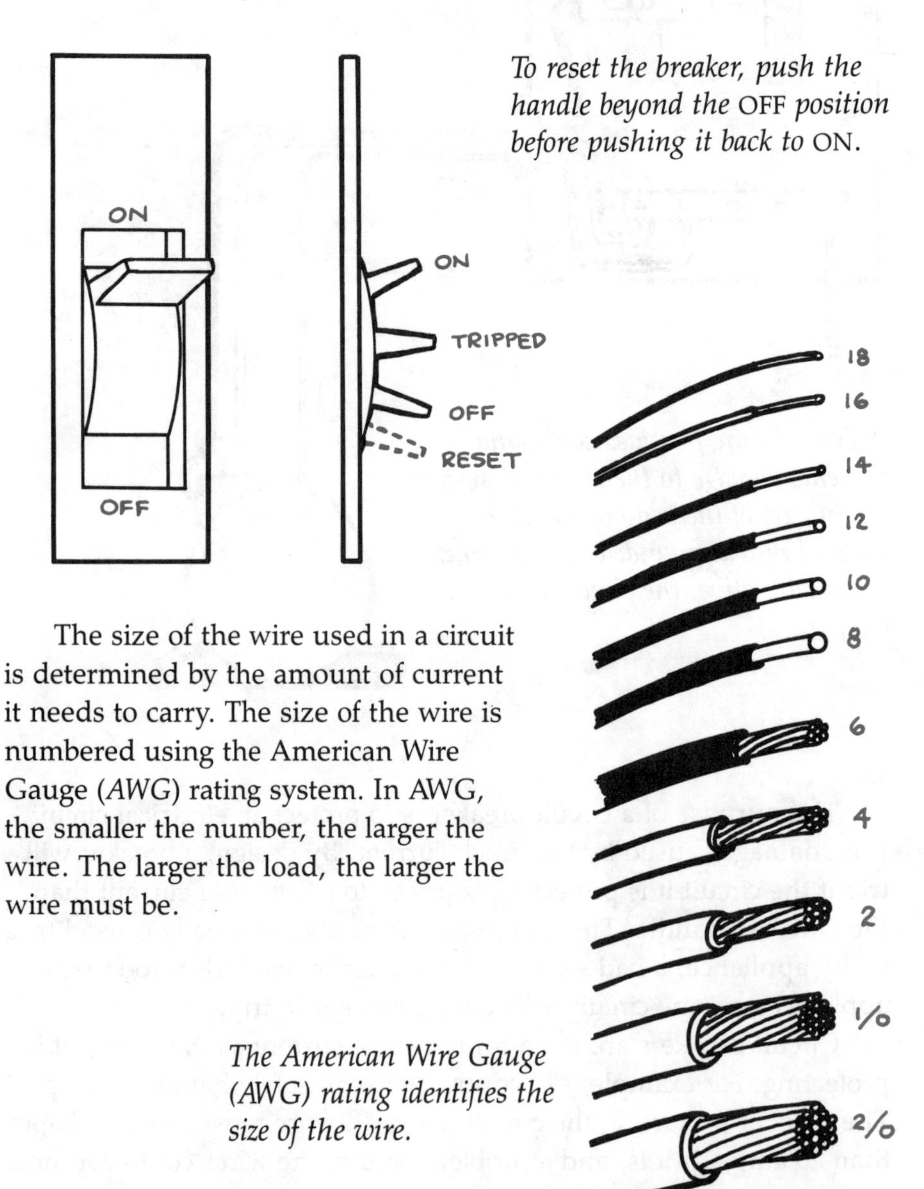


*To reset the breaker, push the handle beyond the* OFF *position before pushing it back to* ON.

The size of the wire used in a circuit is determined by the amount of current it needs to carry. The size of the wire is numbered using the American Wire Gauge (*AWG*) rating system. In AWG, the smaller the number, the larger the wire. The larger the load, the larger the wire must be.

*The American Wire Gauge (AWG) rating identifies the size of the wire.*

A circuit supplying power to wall outlets might use #12 wire, while one feeding a clothes dryer might require #8 or #6. Nonmetallic sheathed cable, sometimes called Romex (the manufacturer's trade name), commonly is used in residential wiring.

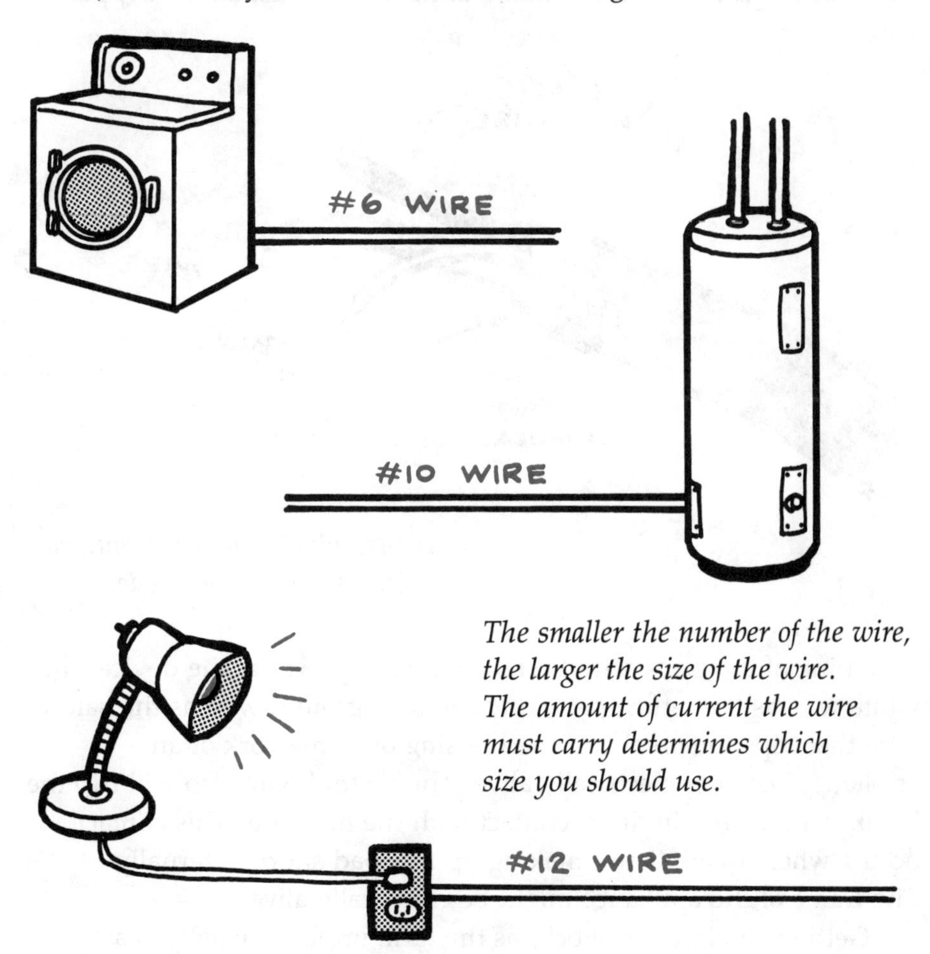


*The smaller the number of the wire, the larger the size of the wire. The amount of current the wire must carry determines which size you should use.*

Cables are identified by the number of wires in the cable. For example, "12/2 with ground" means the cable is made up of two #12 wires and a ground wire. The Code states that the breaker size for copper cables must not be greater than 15 amps for #14 cable, 20 amps for #12, and 30 amps for #10. Although the Code allows #14 cable to be used in some cases, most homes currently being built use #12 as the smallest cable.

To identify the individual wires in the cable, a standard color code is used. The hot wire is black. If a cable has two hot wires, one is black, the other red. The neutral wire is white or gray. The ground wire can be green, but it is usually uninsulated—just a bare copper wire.

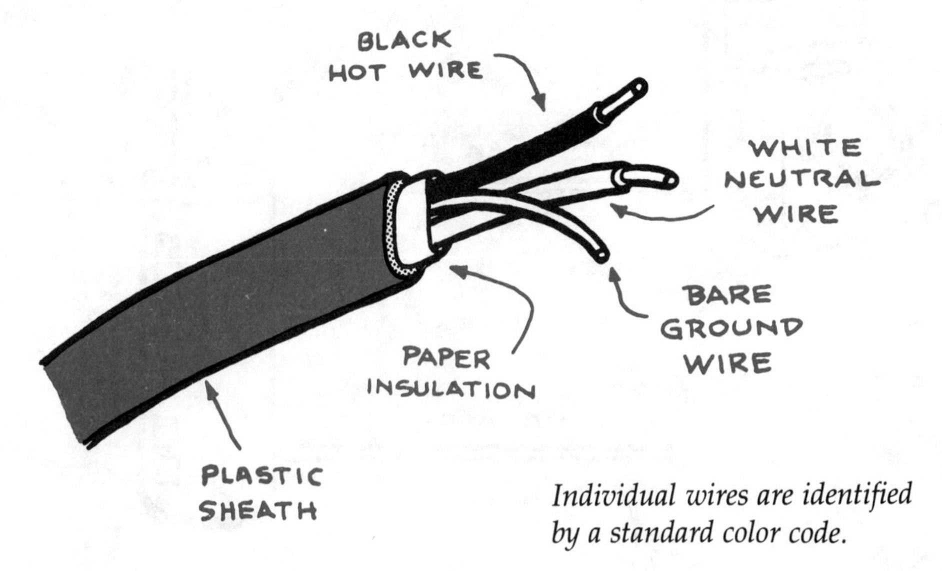


*Individual wires are identified by a standard color code.*

Think of the black wire as delivering the power to the device, the white wire as the return path, and the bare ground wire as the safety wire that trips the breaker if the housing or framework of an appliance becomes electrically alive. The neutral wire also will trip the breaker if it comes in direct contact with the hot wire. This contact occurs when an appliance making up the load shorts internally. The frame of the appliance might be electrically alive.

Getting an electrical shock, as the term implies, usually is a stunning surprise. It can be painful and, in a few cases, fatal. Just as often, injury is caused by involuntary reflexes such as jerking a hand away, falling from a ladder, or some other uncontrollable reaction. Keep in mind, however, that conditions must be exactly right for you to get an electrical shock. Simply touching a wire won't necessarily do it. Remember that current must flow in a continuous path from its source through some device, or load, and then back to the source.

If you happen to become the link in a live circuit, you will receive a shock. You are making the connection that completes the circuit. To receive a shock, you must be touching the live wire and simultaneously come in contact with a grounded object or another live wire. Electricity does not need to flow along wires to make the return trip to its source.

Electricity can return to the source through any conductor, including the human body, that comes in contact with the earth. At first, this situation might seem remote, but remember that water is a good conductor. If you are taking a bath or swimming in a pool, touching any metal water pipes or faucets, or standing on the ground or on a damp concrete garage floor, you become a conductor to ground. Now all you have to do is touch a hot wire or come in contact with a live circuit, and you most certainly will receive a shock. To avoid this situation, observe one of the most important rules for the do-it-yourselfer: Never work on any live circuit, fixture, or appliance.

Before reaching for a screwdriver, go to the service panel and kill the circuit. To ensure that you have killed the right circuit, turn on an overhead light or plug a lamp into a wall outlet before turning off the breaker. Keep in mind that the wall outlets are normally not on the same circuit as the ceiling lights.

If you still have any doubts about the power being off, turn off the main breaker. With the main breaker off, look at the electric meter. The thin wheel, or disk, below the dials should not be turning. At times, however, you will need the power on to perform tests.

If you encounter a situation where someone has been shocked and is still part of the live circuit, never touch the person with your bare hands. First, try to disconnect the power. If you cannot, use some type of insulated device, such as a coat or broom, to remove the victim from the circuit. Then keep the victim warm while you call for help. Artificial respiration might be necessary.

Spend some time familiarizing yourself with how your particular home is wired. If the circuits in the service panel are not labeled, you might consider identifying and labeling them for future reference.

Never be in a hurry. Work slowly and stop and think. Understand what you are doing, and you'll be surprised just how easy it is to work with electricity. Now that the hard, scary stuff is out of the way, let's go to basic tools and troubleshooting.

CHAPTER TWO

# Basic Tools & Troubleshooting

A basic set of tools is usually a necessity for any homeowner. Keep tools together in a place you can find them when you need them. When shopping for tools, remember to buy quality. A poorly made tool is never a bargain. Electrical supply stores have a wide, and often bewildering, range of tools and test equipment. You'll need only a few special tools for most jobs.

You'll need some sort of testing device to check for voltage and *continuity*. An inexpensive tool to check for voltage is the neon voltage tester. If you touch one probe to a live wire or terminal and the other to the neutral or ground and the lamp lights, voltage is present. A continuity check determines if the circuit contains any breaks. A variety of continuity testers are available. They are battery-operated and use a light or buzzer to indicate continuity. These testers are handy to locate breaks in wires. For example, an extension cord can be tested by connecting the alligator clip to one of the prongs on one end of the cord and touching the probe to the other end. If you have the right wire, and the wire is intact, the bulb in the tester lights.

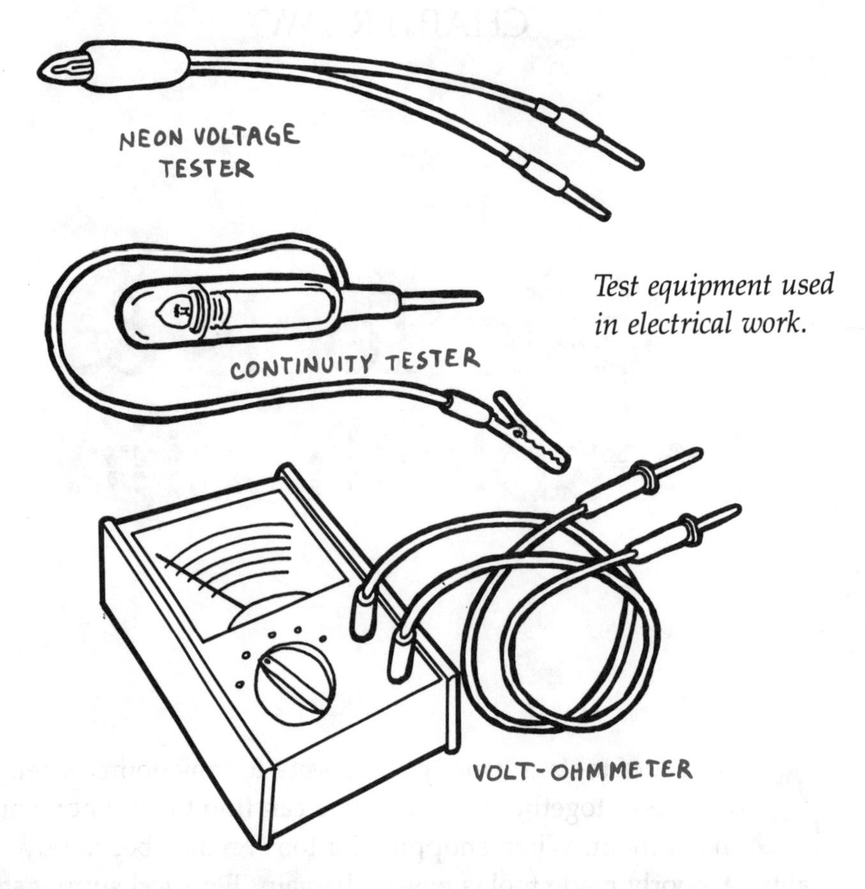


*Test equipment used in electrical work.*

An inexpensive volt-ohmmeter, available at any electronics supply store, checks for both voltage and continuity. This meter also is useful for troubleshooting faulty appliances. When using a volt-ohmmeter to check for continuity, always make sure the meter is switched to the R×1 scale. When testing for voltage, switch to the appropriate voltage scale (usually ac).

When using any test equipment to check for voltage, work slowly and carefully. Don't touch any metal. Keep your hands well back on the insulated areas of the probes. Touch the end of the probes only to the wires or terminals to be tested.

You probably already have a screwdriver or two. One with a standard slot blade about 3/8 inch wide fits most terminal screws. Make sure you use a blade with the correct thickness for the size of the slot in the screw. Another point to remember is that the screwdriver tip should be squared. Screwdrivers are sometimes abused and used for chisels. If the tip becomes rounded, it is no longer a screwdriver. Use a file to keep the tip squared off.

Medium- and small-sized Phillips screwdrivers are useful for cross-slotted screws. Although not common in electrical work, cross-slotted screw heads are found on a number of appliances.

You also need diagonal wire cutters, commonly called "dikes," for cutting wires up to #6. Needle-nose pliers are useful to form a loop in the end of a wire when connecting a wire to a screw terminal.

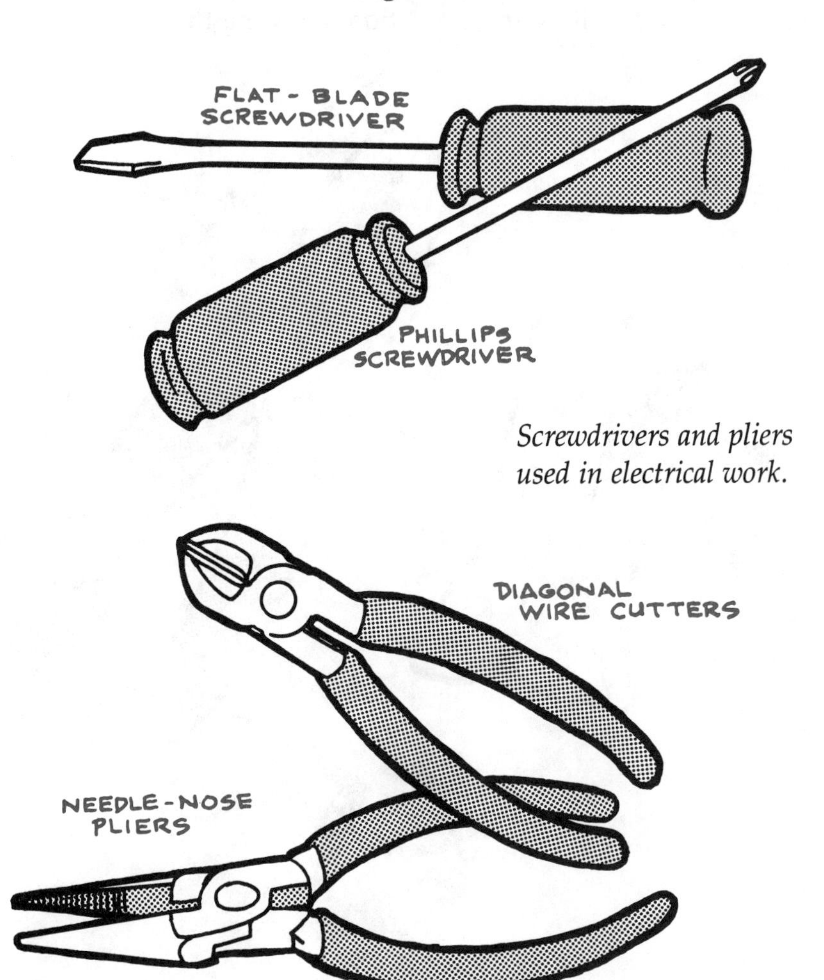


*Screwdrivers and pliers used in electrical work.*

A wire stripper is used to strip the insulation from a wire. A variety of forms are available. The simplest is a two-piece scissor-type arrangement. You might need a little practice before you can strip the insulation without nicking the wire, but you should be able to master it after a few tries.

A cable ripper is an inexpensive tool that speeds up the job of stripping the outer insulation from the flat, two-wire cable used in most homes. This simple tool has a small, triangular blade positioned in such a way that when the tool is pressed around a cable and pulled, it makes a slit in the insulation, allowing the outer cover to be easily peeled.

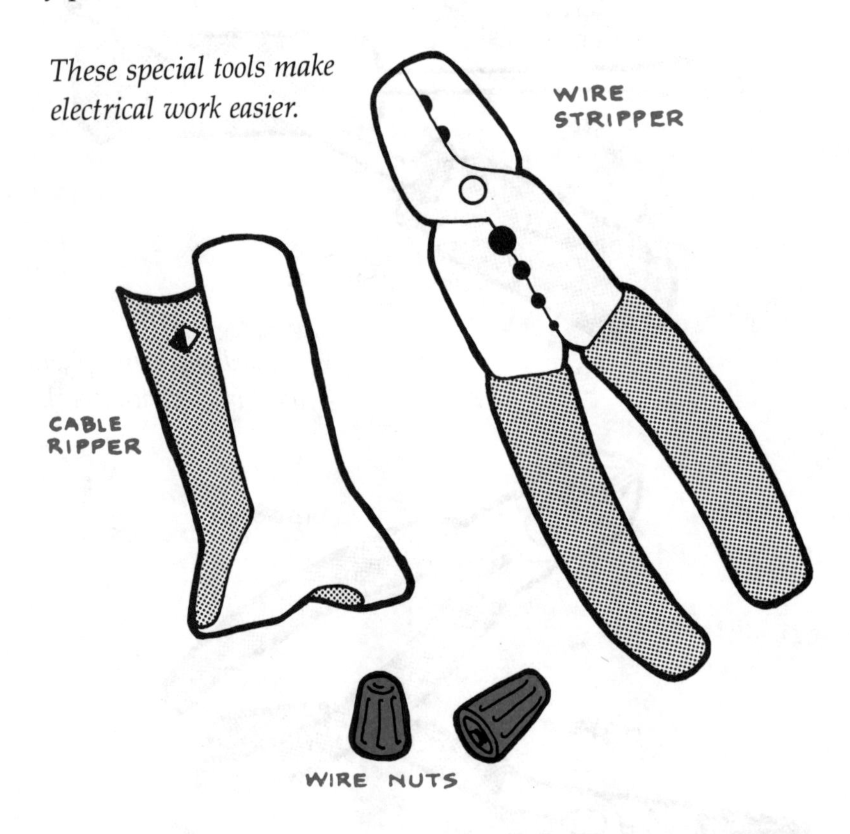


*These special tools make electrical work easier.*

You should be able to find a variety of wire connectors that provide good, solid connections without using solder. Connector size is determined by the size of the wire being used. Sometimes plastic electrical tape is used as an insulating material to cover splices and reinforce wire connections.

Normally, the first sign of a problem is when something doesn't work. The toaster won't heat or the television doesn't come on. A check at the service panel usually reveals a tripped breaker. The most common reason a circuit fails is a *short circuit* or overload. A short circuit means that a shorter path to ground was provided instead of the desired circuit. For example, a hot wire might come in contact with a neutral or ground wire. When this happens, the current suddenly increases beyond the capacity of the breaker, and the circuit fails.

Faulty appliance cords and old extension cords are a common source of trouble. Frayed or worn insulation allows a bare wire to touch another bare wire or metal housing, producing a short circuit. Because most problems occur in flexible cords, plugs, or appliances, you usually can find a short circuit in a matter of minutes. Look for blackened areas on outlets or charred or frayed appliance cords plugged into the dead circuit. Unplug the faulty cord or appliance and reset the breaker.

If you find no visible evidence of any problem, unplug all the appliances and turn off all wall switches on the dead circuit. Then reset the breaker. If the breaker trips again, make sure it is off. Remove the covers and pull out the switches or receptacles one at a time. Look for charred insulation or burn marks caused by a bare wire touching the back of a metal box. The switch or the receptacle itself might even be broken.

If the breaker doesn't trip, turn on the wall switches one at a time until the breaker trips. The problem is probably in the fixture controlled by the switch or the switch itself. Make sure the circuit is dead, and then inspect the fixture and switch. Look for the obvious black marks, charred wires, or burned wire connections.

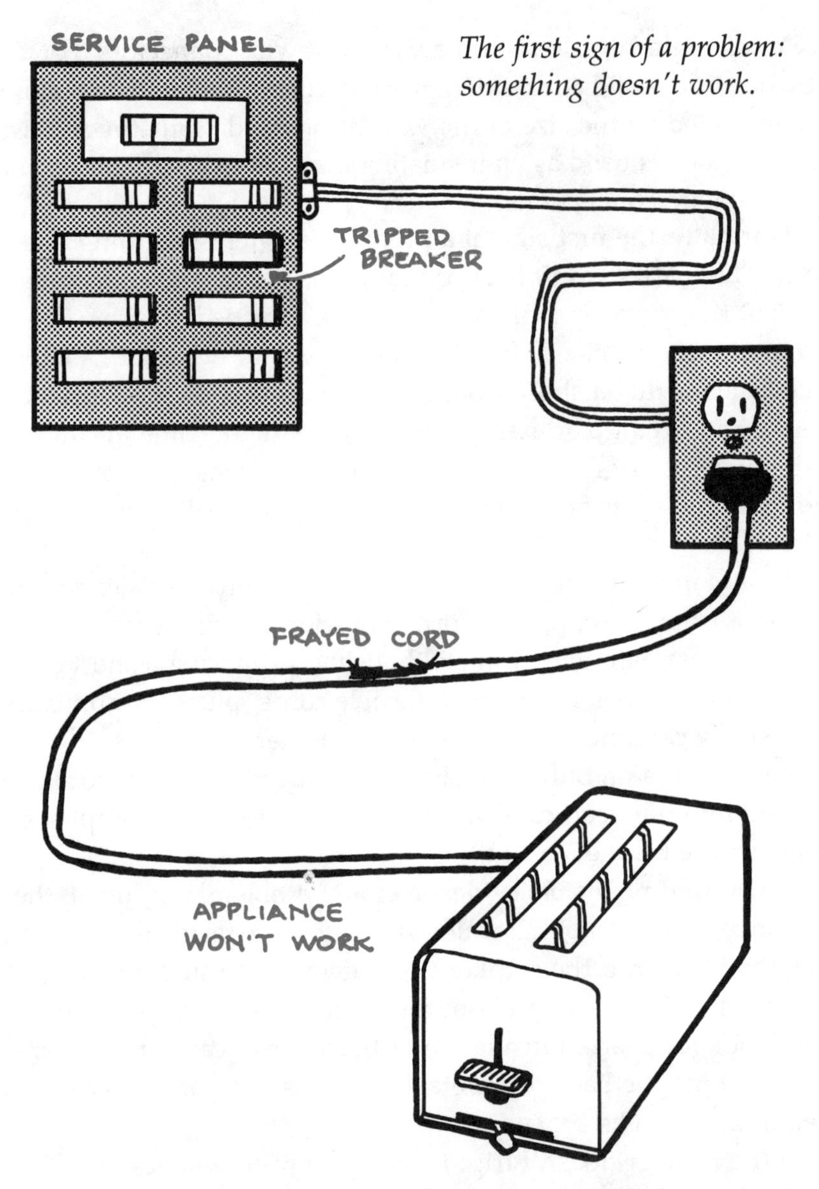


*The first sign of a problem: something doesn't work.*

If the breaker doesn't trip and all the switches are on, then the problem is in one of the appliances you unplugged earlier. Plug in and turn on one appliance at a time until the circuit fails. If the breaker trips when the appliance is plugged in, inspect the plug and cord. However, if the circuit fails only when the appliance is turned on, the switch or the appliance itself is defective.

CHAPTER THREE

# Stripping Wire & Making Connections

Almost all home wiring projects involve stripping insulation from a wire and making some type of electrical connection. These are basic, but very important, steps in wiring. Loose connections can overheat and even cause fires. When connecting wires or attaching them to screw terminals, take a little time to make good, solid connections.

## Tools & Materials

- ☐ Pocket knife
- ☐ Cable ripper
- ☐ Dikes
- ☐ Wire stripper
- ☐ Needle-nose pliers
- ☐ Screwdriver
- ☐ Wire nut

**Step 3-1. Cutting the outer cover.**
Before any connections can be made, the outer covering of the cable must be removed. Make a cut down the center of the cable. You can use a knife, but you risk cutting the insulation on the wires. To use a cable ripper, place the tool over the cable, squeeze the tool, and pull it toward the end of the cable.

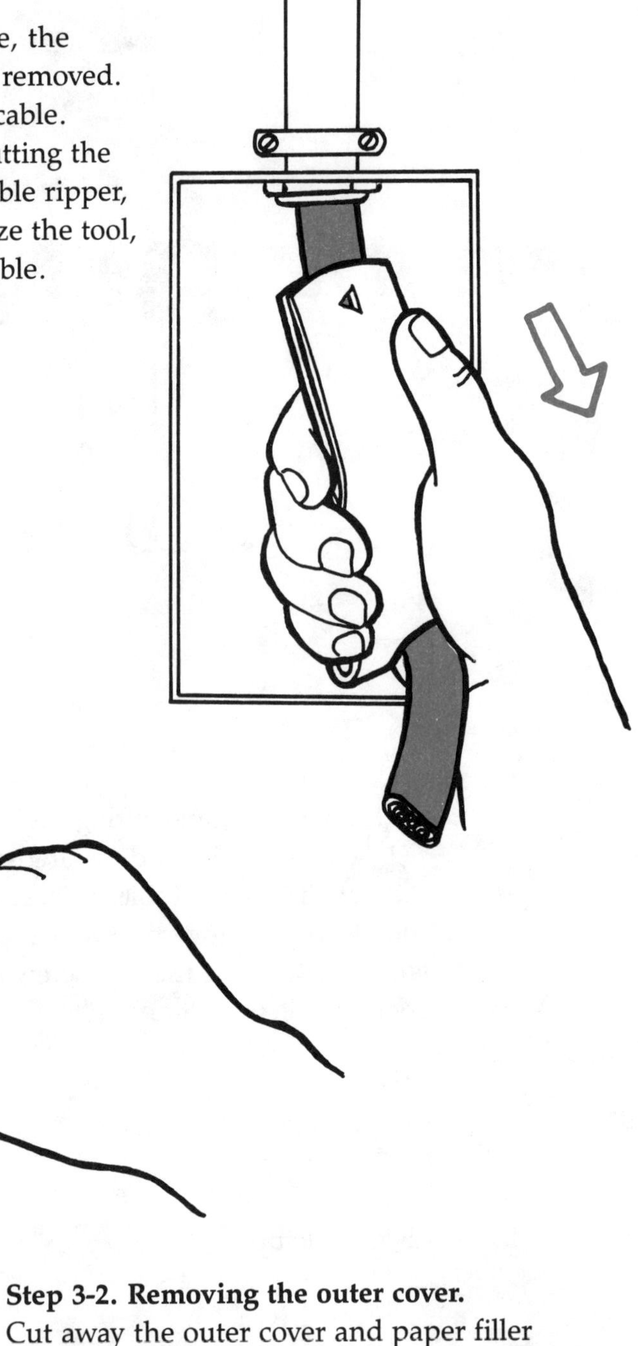

**Step 3-2. Removing the outer cover.**
Cut away the outer cover and paper filler with a knife or diagonal wire cutters (dikes).

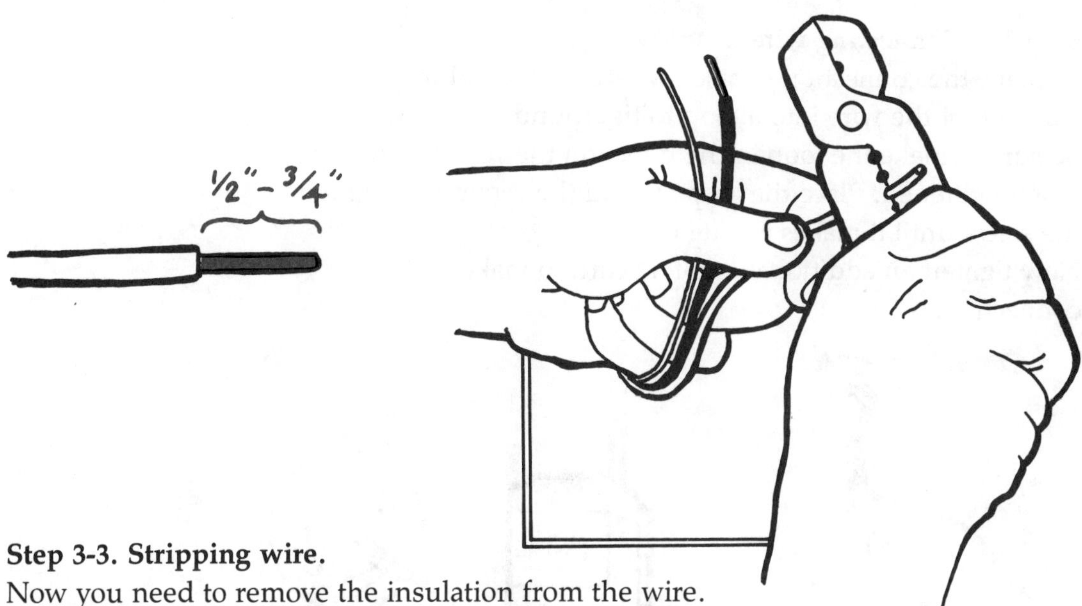


**Step 3-3. Stripping wire.**
Now you need to remove the insulation from the wire. Depending on the connection, you'll want to remove between 1/2 to 3/4 inch of insulation. If you use a wire stripper with notched jaws, simply match the notch with the wire size, place the wire in the notch, and squeeze the handle. A gentle pull outward will strip the insulation from the wire. Other wire strippers might not have a stop that prevents cutting into the wire, so take care when removing insulation.

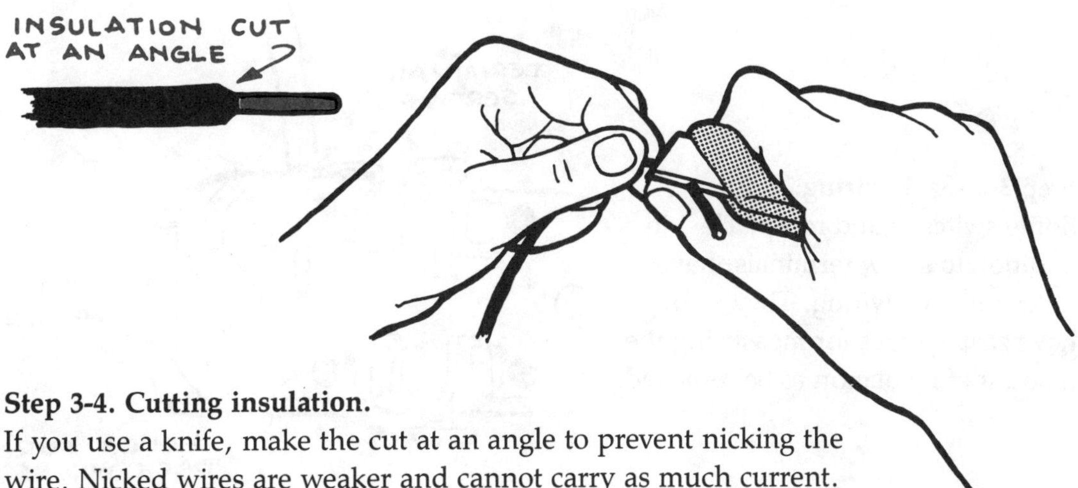


**Step 3-4. Cutting insulation.**
If you use a knife, make the cut at an angle to prevent nicking the wire. Nicked wires are weaker and cannot carry as much current.

**Step 3-5. Connecting wire to screw.**
To make the connection to a screw terminal, bend the bare end of the wire into a loop to fit around the screw. Be sure to make the loop in the direction the screw is turned when tightened. Place the loop around the screw and tighten the screw until it makes a snug contact. Now tighten an additional one-half turn to make a firm connection.

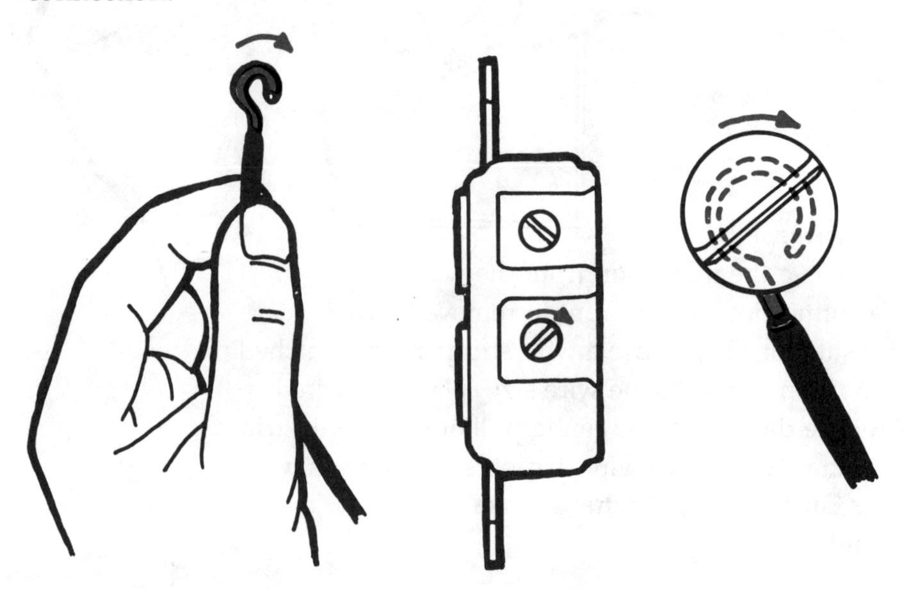

**Step 3-6. Back wiring.**
Some switches and receptacles, in addition to screw terminals, have holes for back wiring. They also have strip gauges for measuring the amount of insulation to be removed.

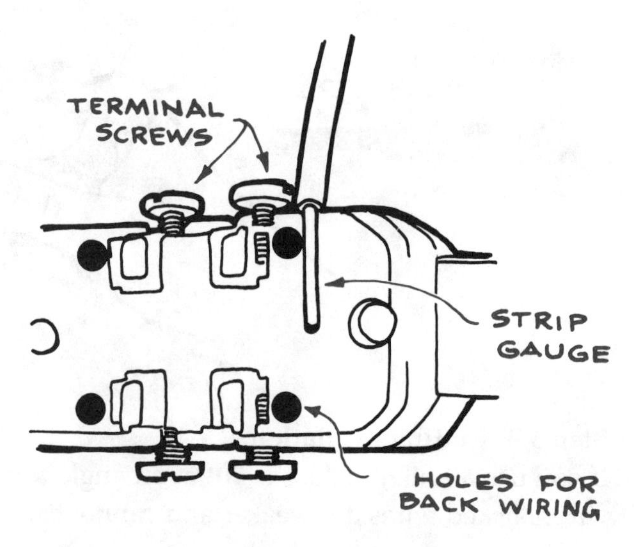

**Step 3-7. Inserting wire.**
After the insulation is removed, insert the bare end of the wire into the proper hole.

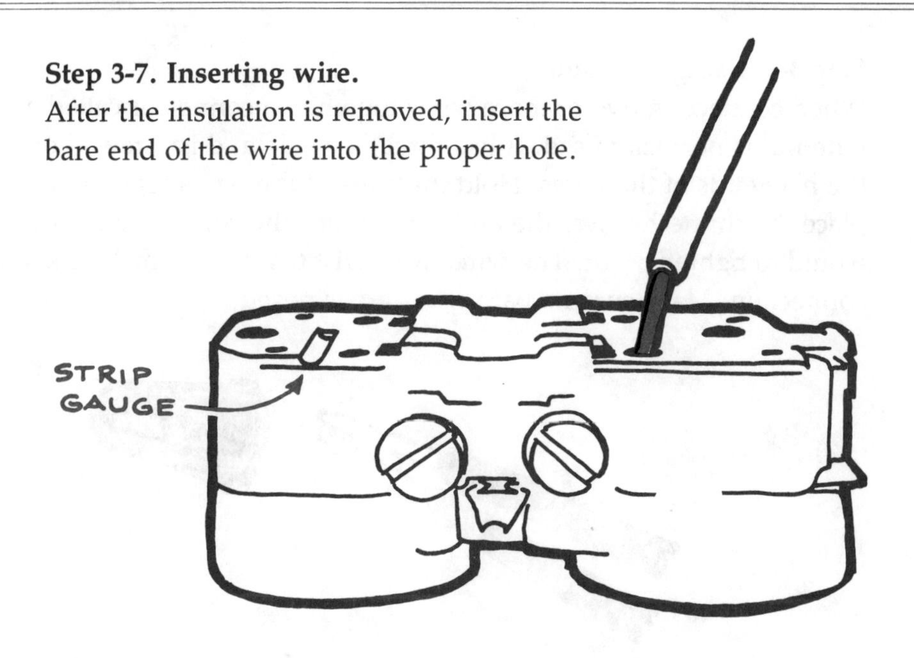


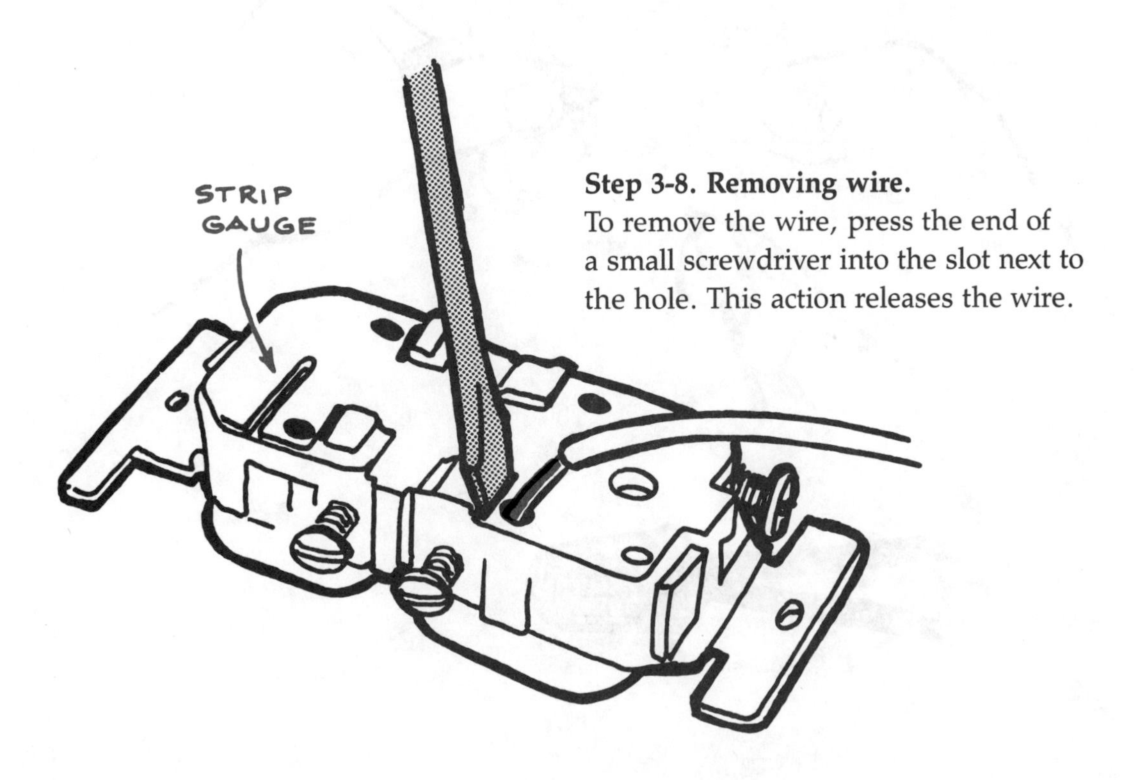


**Step 3-8. Removing wire.**
To remove the wire, press the end of a small screwdriver into the slot next to the hole. This action releases the wire.

**Step 3-9. Using wire nuts.**
When connecting two or more wires together, a connector called a wire nut is normally used. It has threads inside and is screwed on the bare ends of the wires. Hold the ends of the wires together and place the connector over the ends. Now turn the connector as you would to tighten a nut. The wires will twist together to make a solid connection. Make sure no bare wires are exposed.

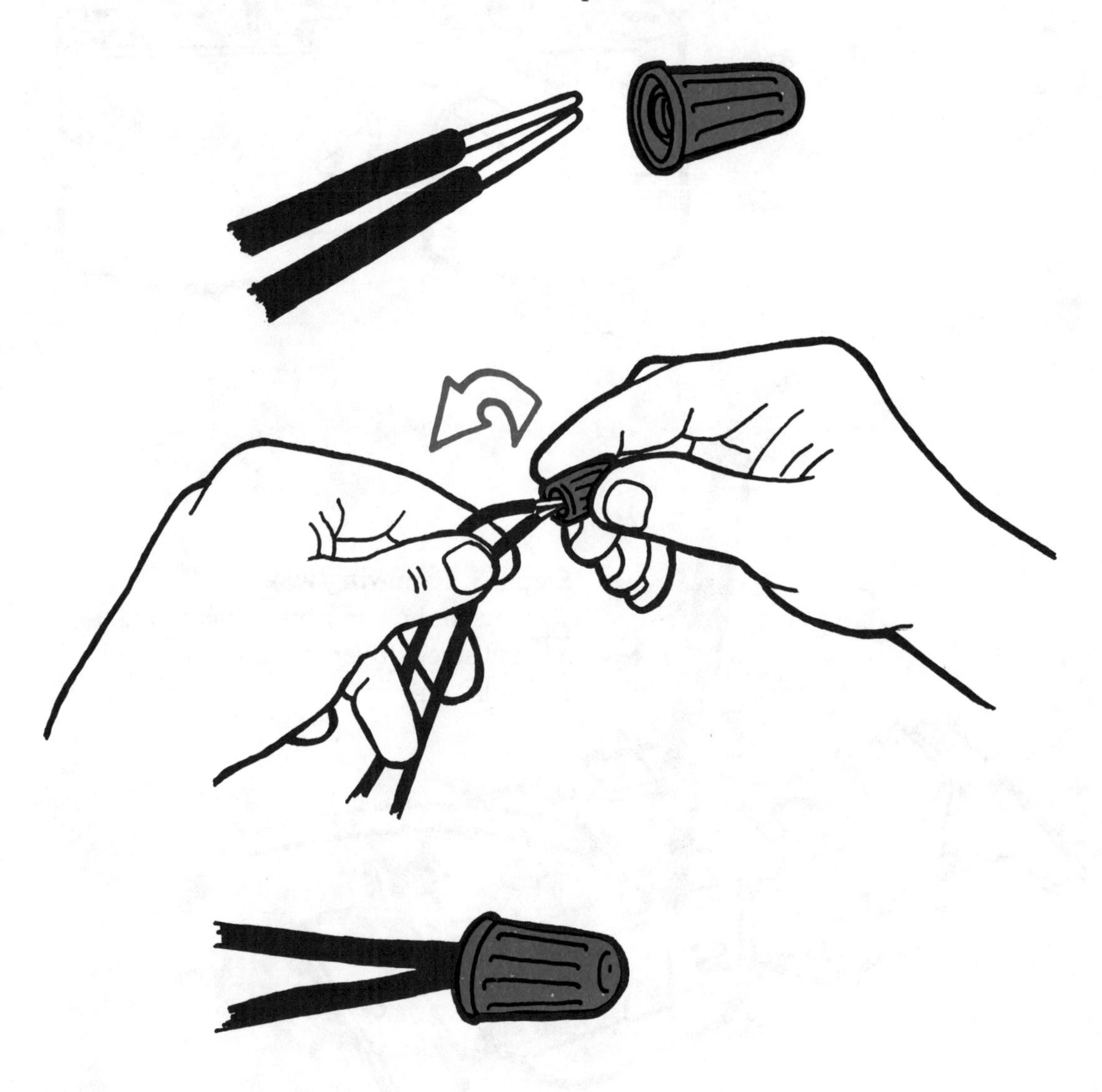

CHAPTER FOUR

# Replacing Switches

Anytime you flip a switch and the light doesn't come on—but you know that the bulb is good and the breaker hasn't tripped—the problem is usually a faulty switch. A switch is just a device that controls the current flow in a circuit. Several types of switches are used in homes. The most common one is the single-pole switch, but you might have three- and four-way switches as well. Switches are mounted in plastic or metal wall boxes and covered with a wall plate. They are very easy to replace.

## Tools & Materials

- ☐ Screwdriver
- ☐ Pocket knife or wire stripper
- ☐ Needle-nose pliers
- ☐ Neon voltage tester or volt-ohmmeter
- ☐ New switch

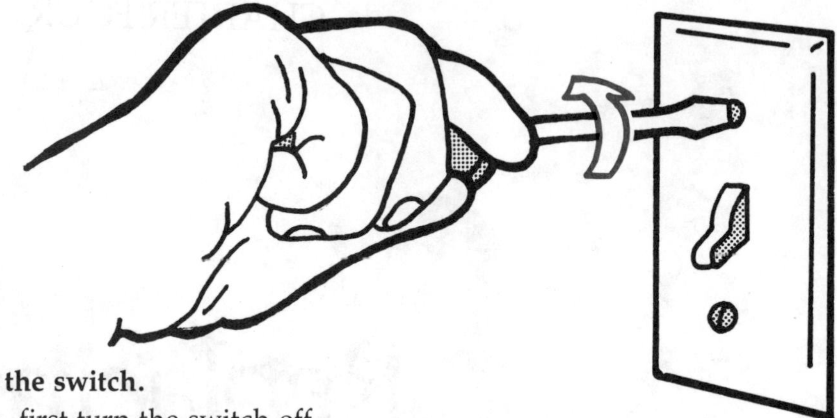

**Step 4-1. Checking the switch.**
To check the switch, first turn the switch off and remove the cover plate. Now touch one of the test probes of a voltage tester to the metal housing of the switch (ground), and the other probe to each of the screw terminals on the switch. The tester lamp should light when the probe touches one of the screws but not the other. If the lamp does not light at all, the switch is not receiving power, and it might not be the problem. If the switch is receiving power, turn it on. If the switch is good, the lamp now should light when the probe touches either screw.

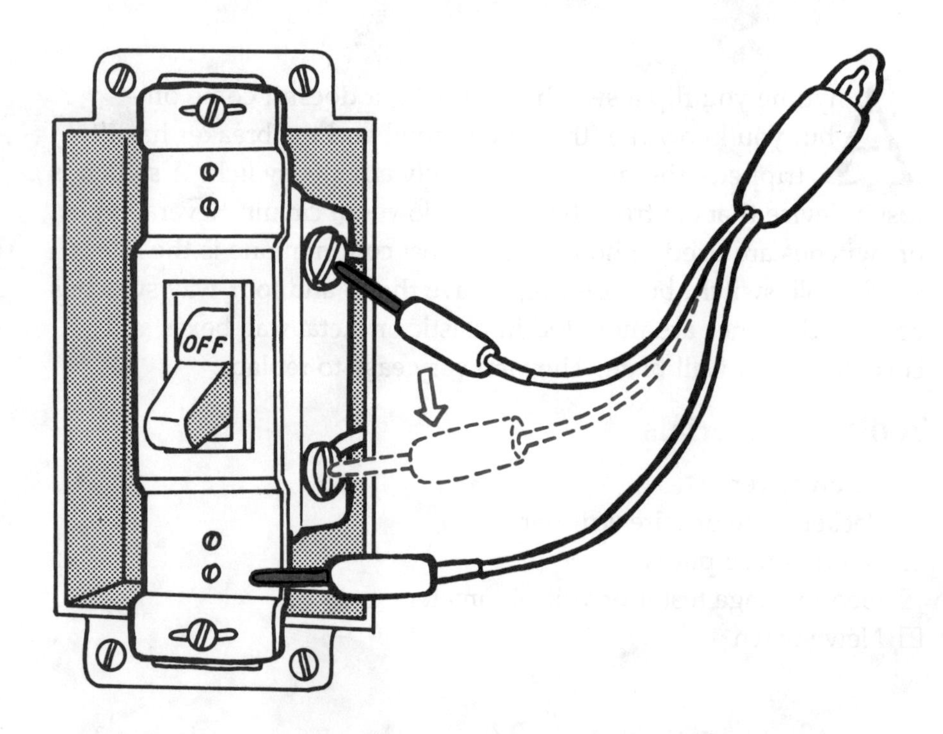

**Step 4-2. Checking an ungrounded switch.**
In some installations, the switch will not be grounded. In this case, touch a probe to each screw terminal. The tester lamp should light with the switch off and go out with the switch on. If just one of the screws has voltage when the switch is on, the switch is probably bad. Go to the service panel and turn off the breaker controlling the circuit. To make sure the circuit is dead, check it again with the tester.

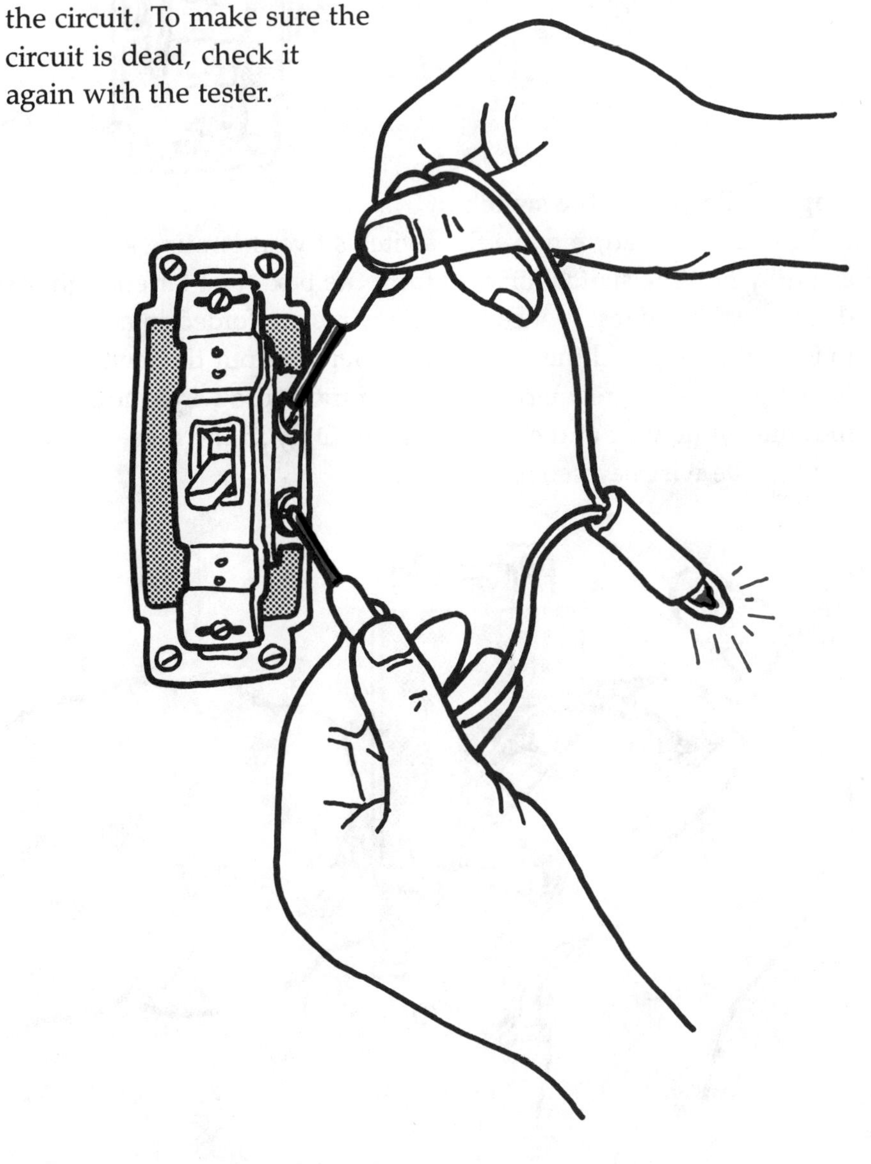

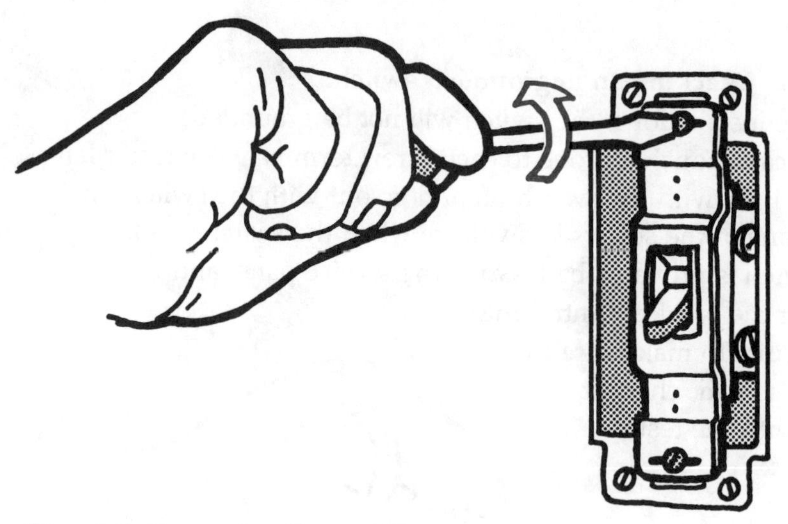

**Step 4-3. Removing the switch.**
If the circuit is dead, remove the switch's two mounting screws, and carefully pull the switch forward from the box—just far enough to disconnect the wires. Notice that the wires are folded in an S-shape to fit in the box. Fold them this way when you put the switch back. Now, loosen the screw terminals and remove the wires. You might find one white wire and one black wire. Only in the case of switches can a white wire be used as a hot wire.

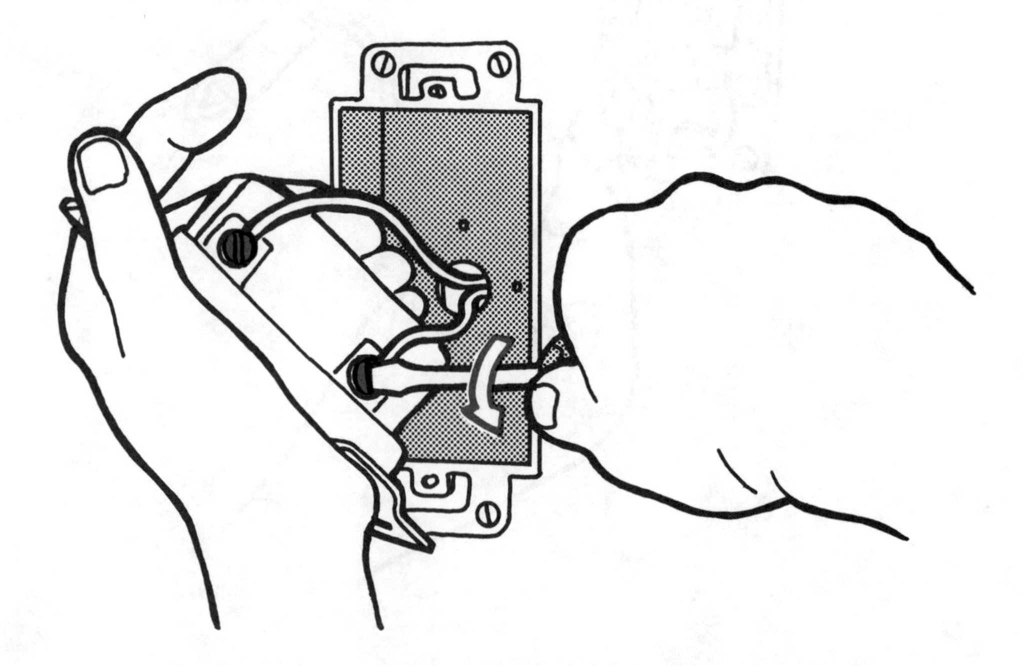

**Step 4-4. Removing a back-wired switch.**
If the switch is back-wired, press the tip of a small screwdriver into the slot next to the hole containing the wire and pull out the wire. Some back-wired switches use the terminal screw to hold the wire. In this case, loosen the screw and remove the wire from the hole.

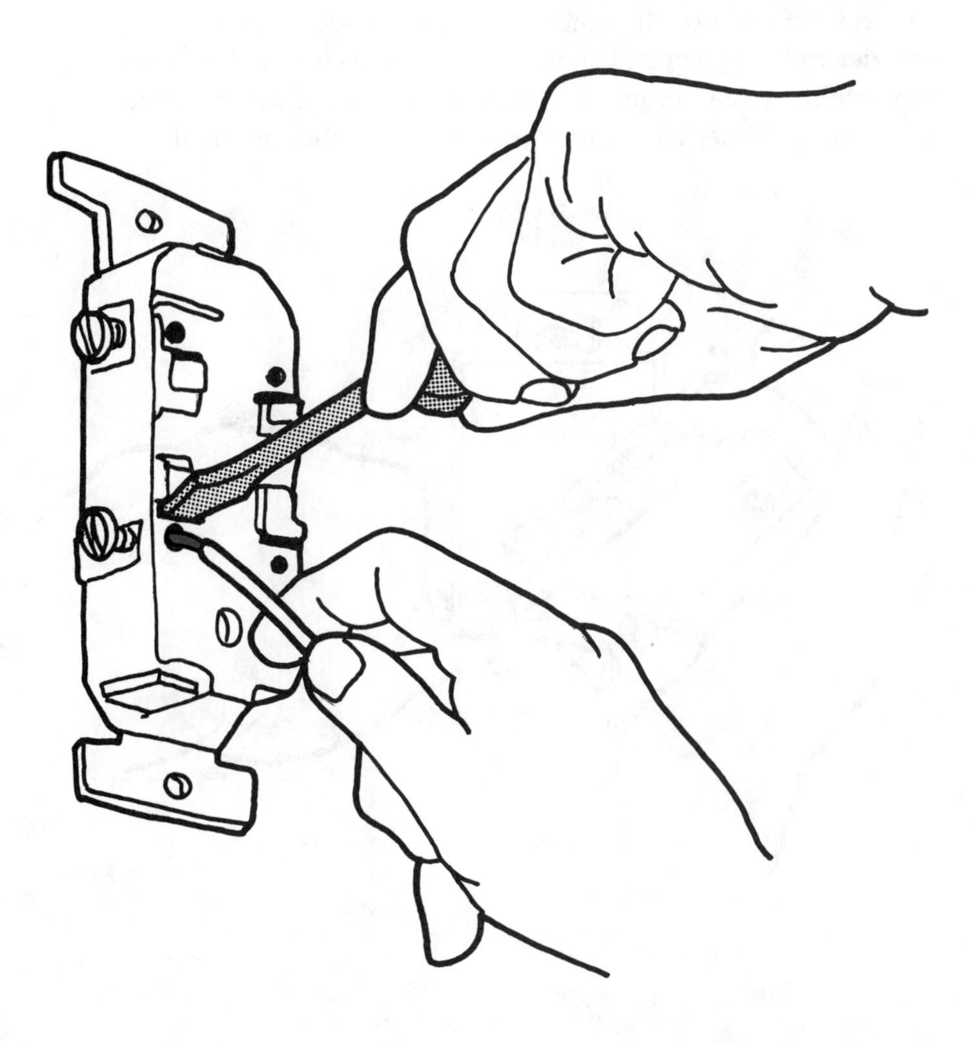

**Step 4-5. Installing a switch.**
To install the new switch, prepare the ends of the wires. Use loops for screw terminals or make the ends straight for back wiring. Loosen the screw terminals, but don't take them all the way out. A switch has a top and bottom. The toggle is down when the switch is off and up when the switch is on. Look at the switch and determine which end should be up. Now connect the wires to the terminal screws or insert them into the back of the switch for back wiring. Either wire can be connected to either terminal.

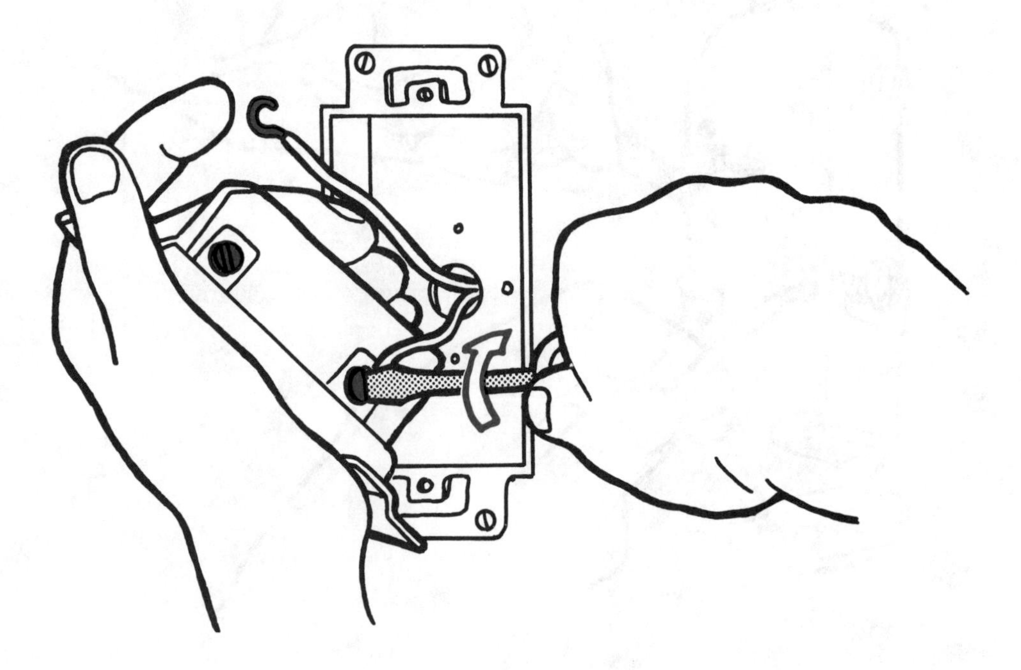

**Step 4-6. Mounting the switch.**
Fold the slack part of the wires as you push the switch back into the box. Install the mounting screws in the box. Notice that the holes in the switch bracket are slotted to allow you to align the switch. Center the switch and tighten the screws.

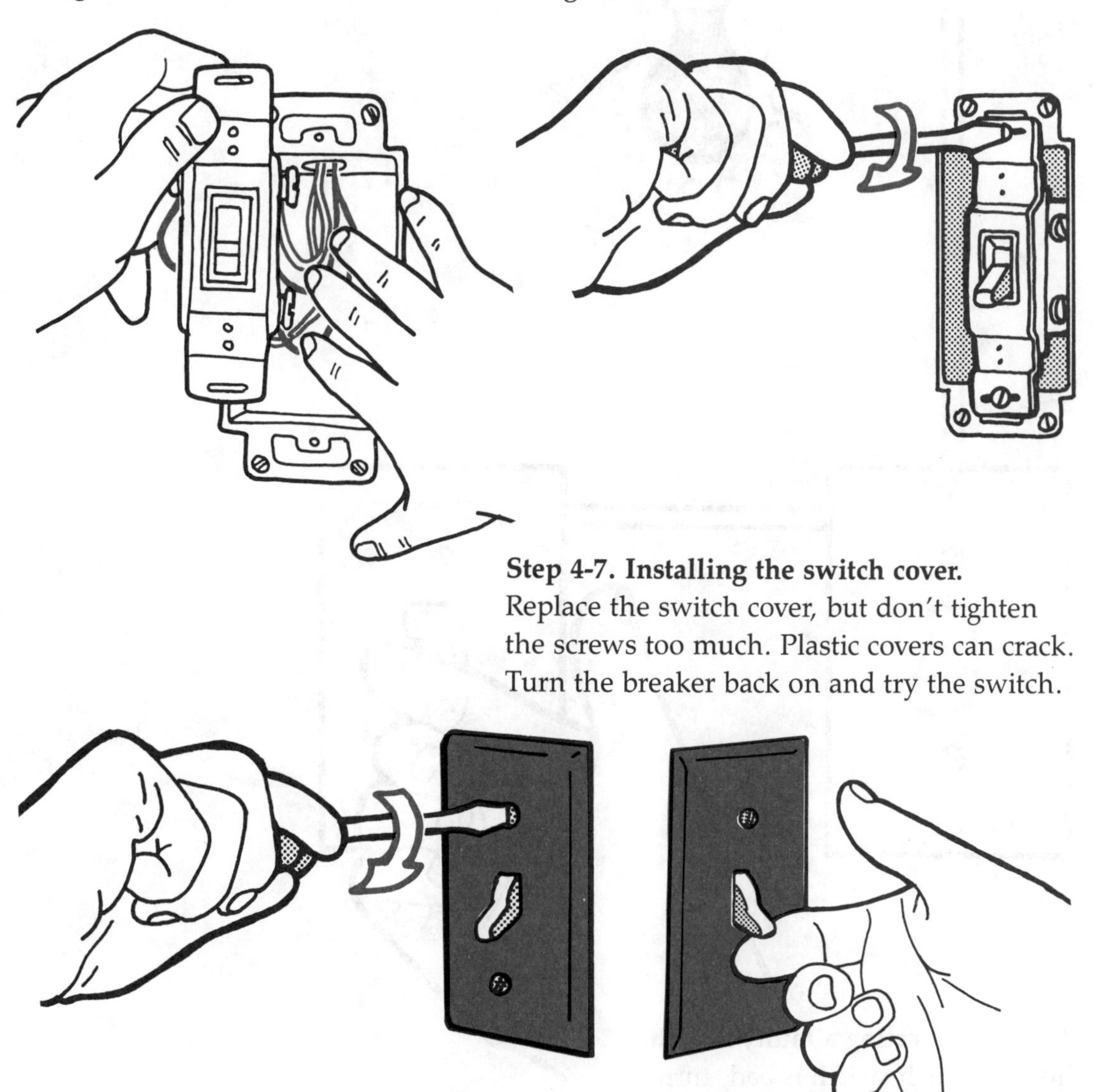

**Step 4-7. Installing the switch cover.**
Replace the switch cover, but don't tighten the screws too much. Plastic covers can crack. Turn the breaker back on and try the switch.

Three-way switches are found where you want to turn a light on or off at two different locations—not three as the term implies. The term *three-way* means that the switch has three terminals for three wires instead of two. The third terminal is called a common terminal. Three-way switches have no ON and OFF markings.

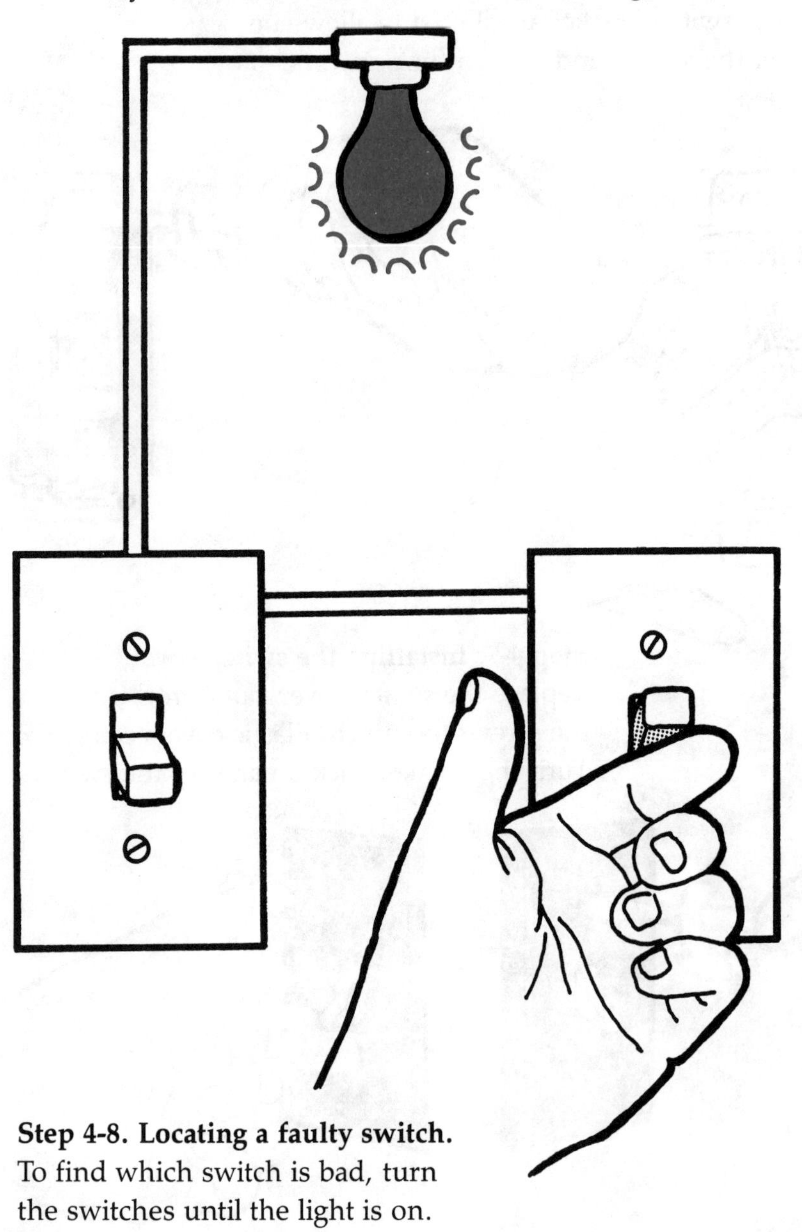

**Step 4-8. Locating a faulty switch.** To find which switch is bad, turn the switches until the light is on.

**Step 4-9. Setting switches to test.**
Now turn each switch to the opposite position to set the switches so that you can make the tests. Remember that the circuit is live. Be careful.

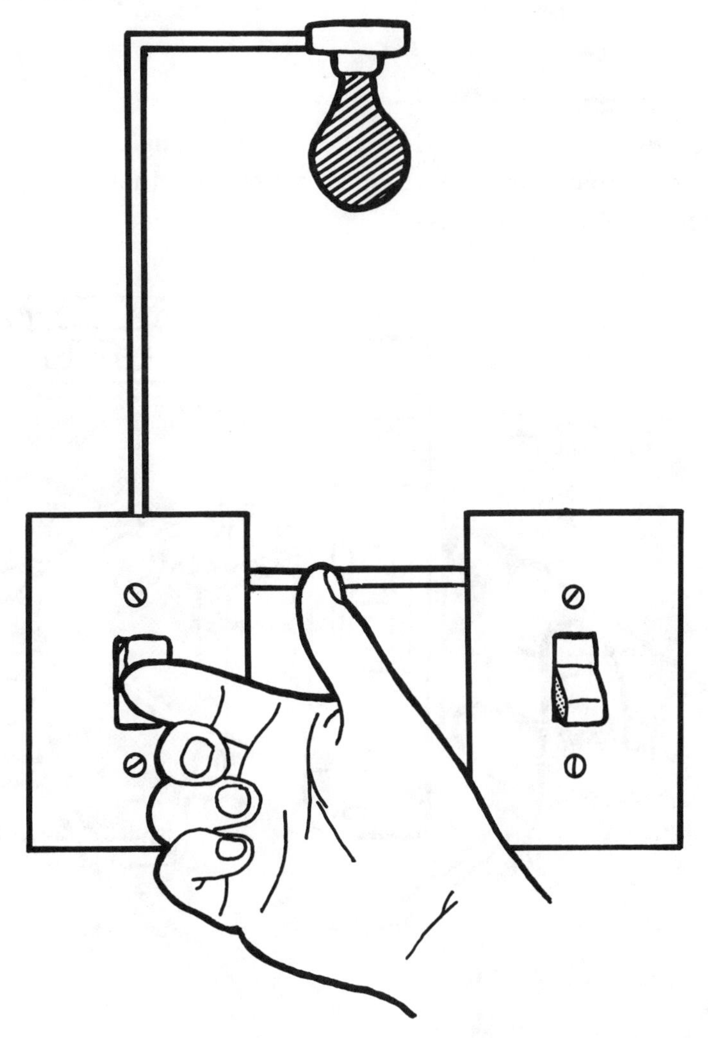

Remove the switch cover from one of the switches. You should see three terminals. If they are not accessible, you'll have to pull the switch from the box to make the test. In this case, turn off the breaker and remove the switch-mounting screws. Pull the switch out far enough to get to the terminals. The wires will hold the switch clear of the box.

**Step 4-10. First test.**
Turn the breaker back on and touch the end of one probe of the voltage tester to one of the terminals. Touch the other probe to one of the other terminals.

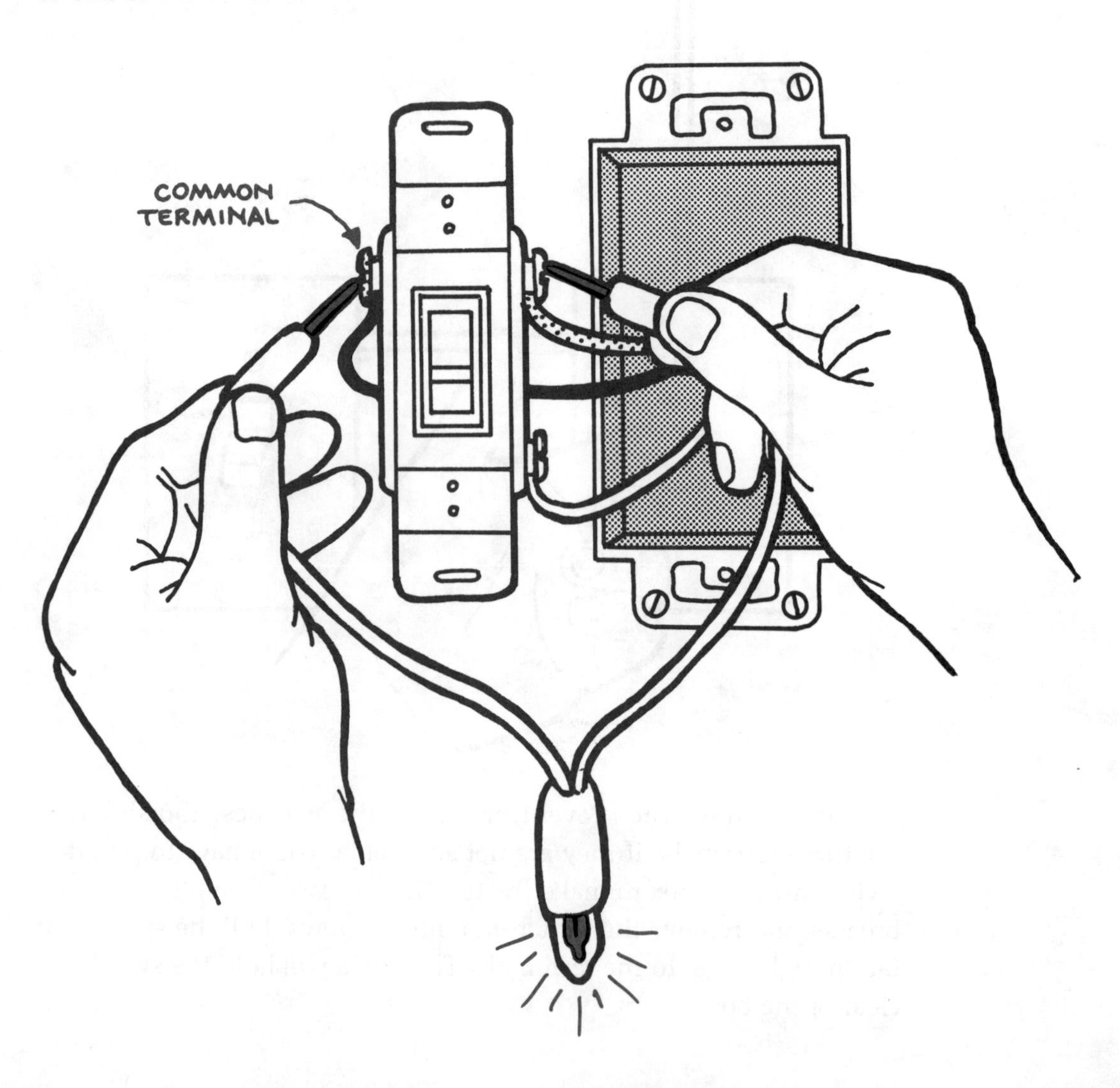

**Step 4-11. Second test.**
Move one of the probes to a different terminal.

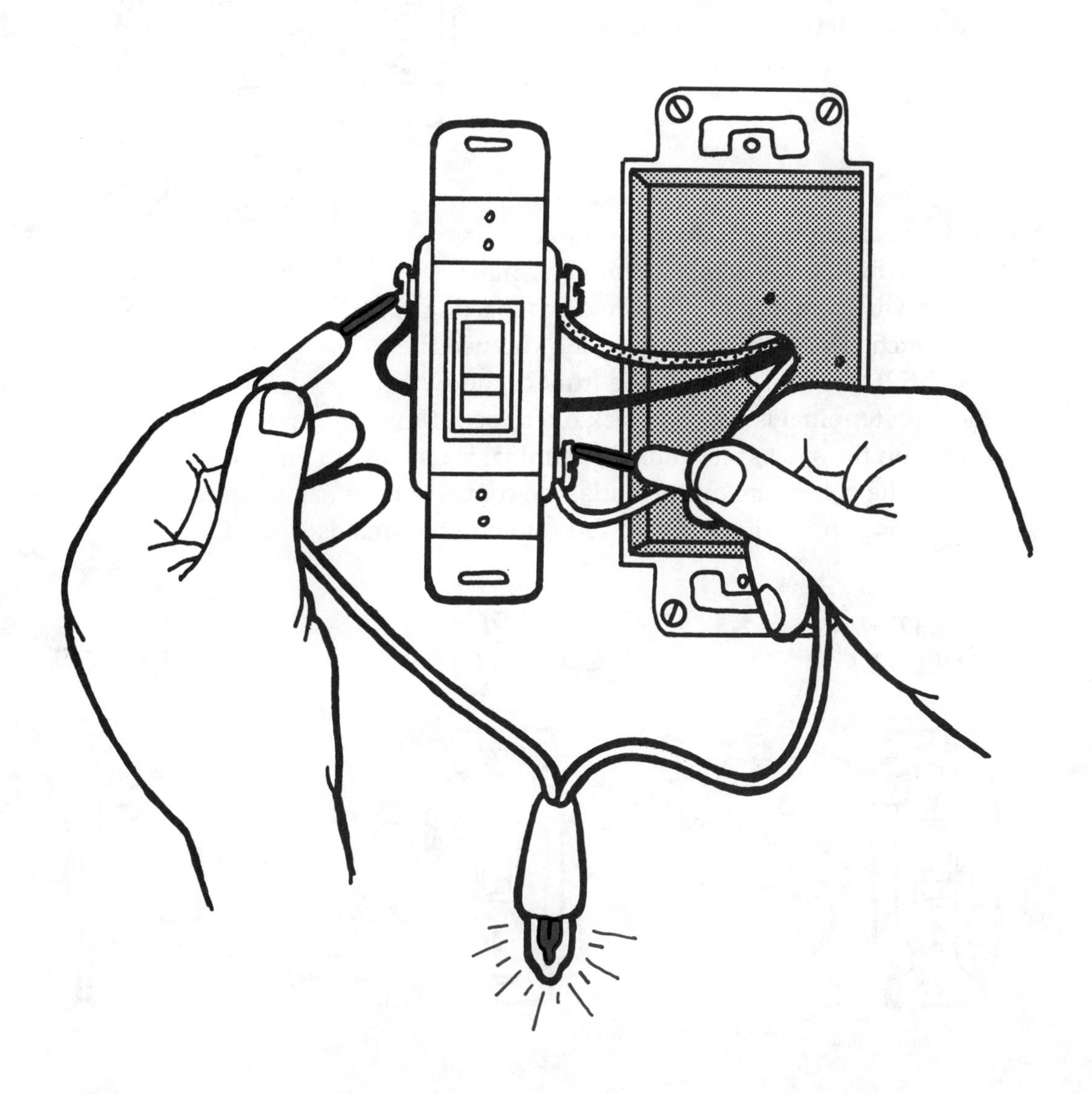

**Step 4-12. Third test.**
Check for voltage between the remaining two terminals. If you found voltage (the test lamp burned) at any one of the combinations, the switch is bad. If the test lamp showed no voltage, the switch is good. ***Caution!*** The test lamp showing no voltage does not mean that no voltage is on the switch. You are only checking for a break in the switch. If the switch checks as good, test the other switch. After you have determined which switch is faulty, turn off the breaker.

**Step 4-13. Locating the common terminal.**
Determine which screw is the common terminal and which wire connects to it before disconnecting the old switch. This wire is the only one you need to get right. If the screws are on the front of the switch, the common terminal is usually black or copper. If the screws are on the sides, the common terminal is usually on one side by itself. If the screws are on the ends, the common terminal is usually on one end by itself. Often the common terminal is labeled.

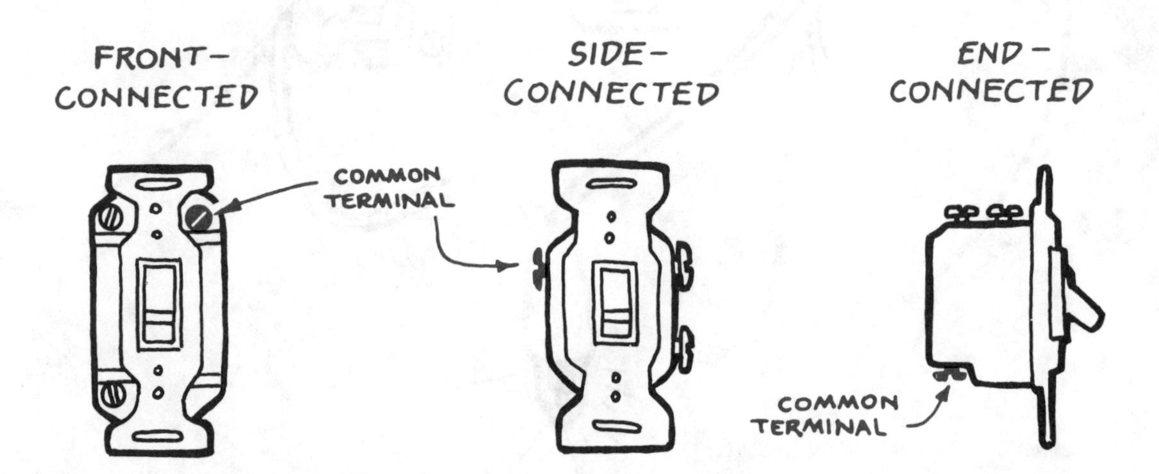

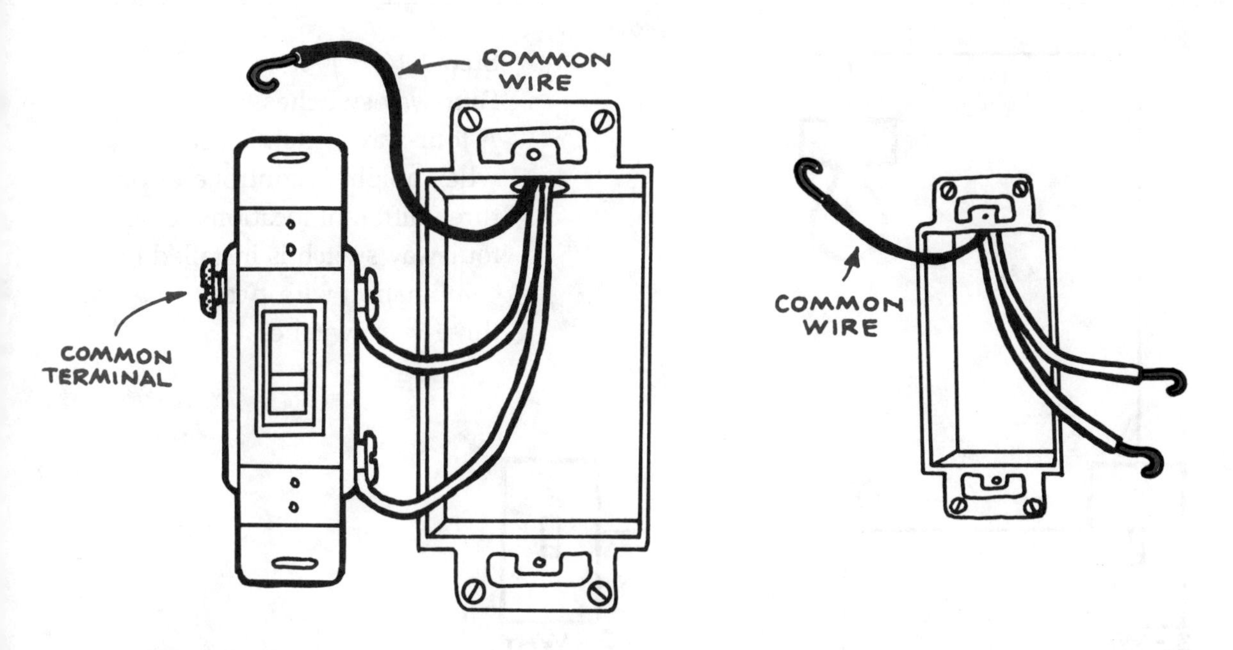


**Step 4-14. Disconnecting wires.**
After you have it located, disconnect the wire from the common terminal and bend the wire away from the switch so you can identify it later. Disconnect the other two wires and bend them to the other side.

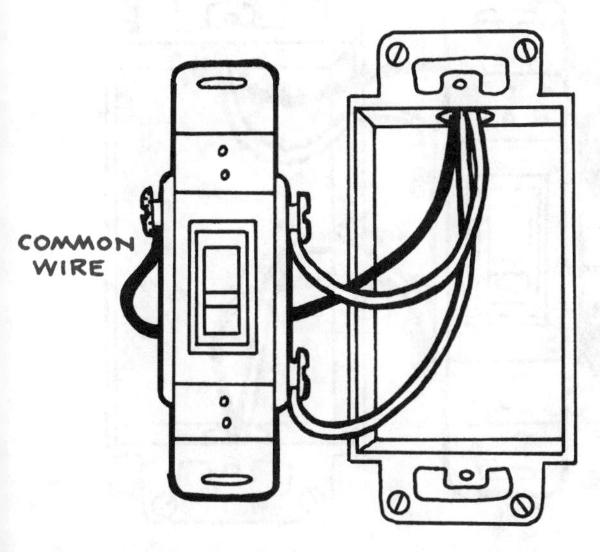


**Step 4-15. Installing a new switch.**
Examine the new switch to locate the common terminal. The box the switch came in might have a wiring diagram. Connect the common wire to the common terminal and connect the other two wires to the remaining terminals. It doesn't matter which of the last two wires are connected to which terminal—as long as the common is connected properly. Fold the wires back into the box, replace the switch-mounting screws, and replace the switch cover. Now turn the breaker back on.

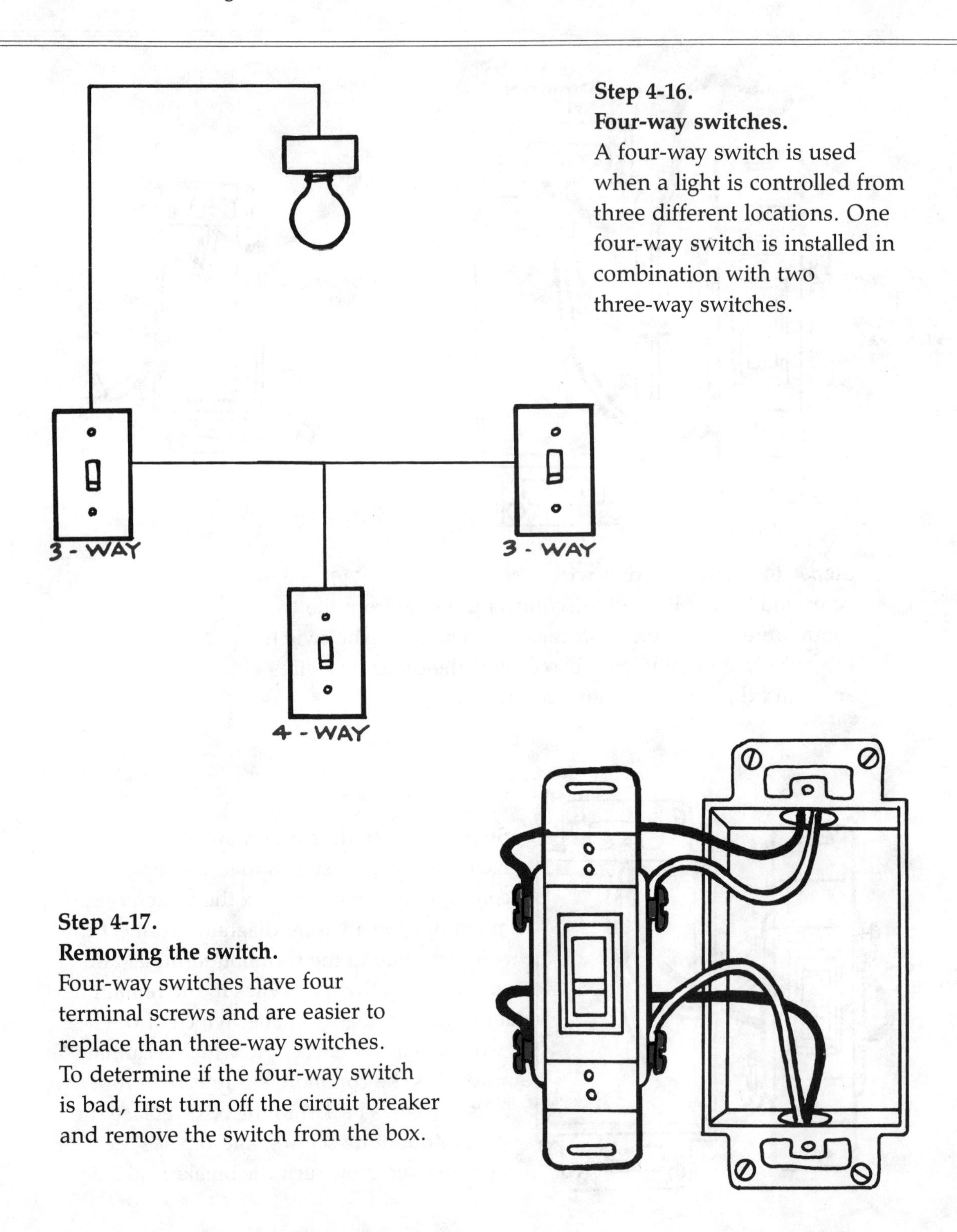


**Step 4-16.**
**Four-way switches.**
A four-way switch is used when a light is controlled from three different locations. One four-way switch is installed in combination with two three-way switches.

**Step 4-17.**
**Removing the switch.**
Four-way switches have four terminal screws and are easier to replace than three-way switches. To determine if the four-way switch is bad, first turn off the circuit breaker and remove the switch from the box.

**Step 4-18. Disconnecting wires.**
Disconnect the wires from the switch. Notice that, in the box, four wires come from two different cables. Temporarily hook the bare end of a wire from one cable to the same-colored wire of the other cable. Connect the remaining two wires the same way. Make sure the wires are clear of the box.

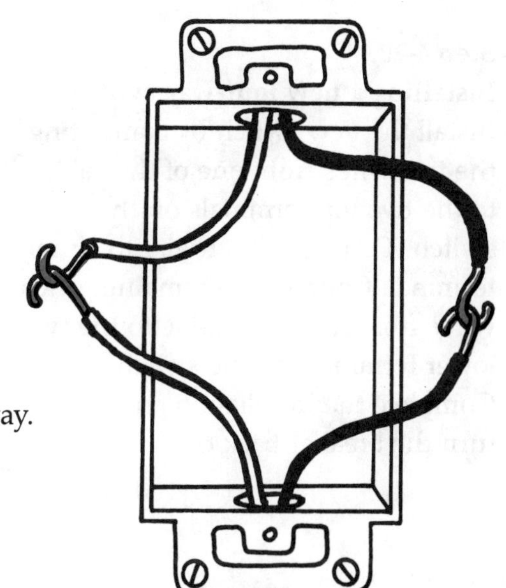

**Step 4-19.**
**Testing for a faulty switch.**
Turn the breaker back on and try the three-way switches. If those switches work correctly, the four-way switch is bad. If they only operate the light in certain positions, the four-way switch is probably good. If the four-way switch is good, turn the breaker back off and reinstall the four-way switch. Check the three-way switches (steps 4-8 through 4-15). If the four-way switch is bad, leave the breaker off and unhook the wires.

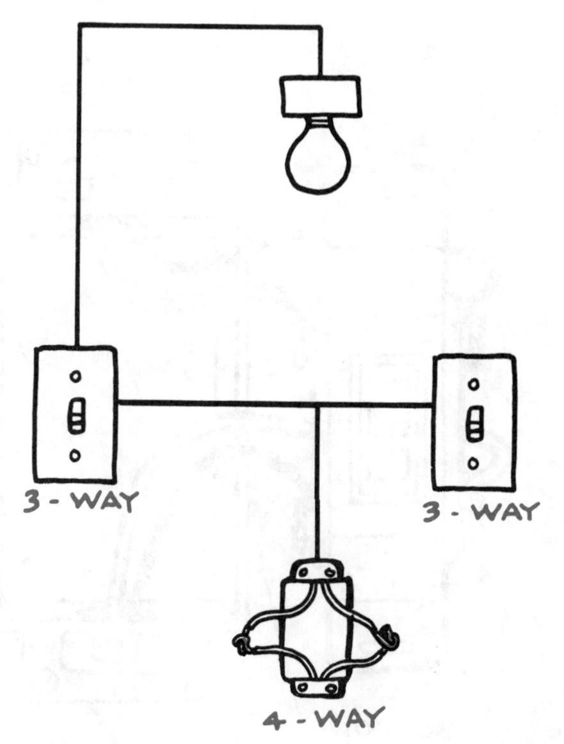

**Step 4-20.**
**Installing a new four-way switch.**
Install the new switch by connecting the two wires from one of the cables to the two top terminals on the new switch. Connect either wire to either terminal. Connect the remaining two wires from the other cable to the two lower terminals on the switch. Complete the installation and turn the breaker back on.

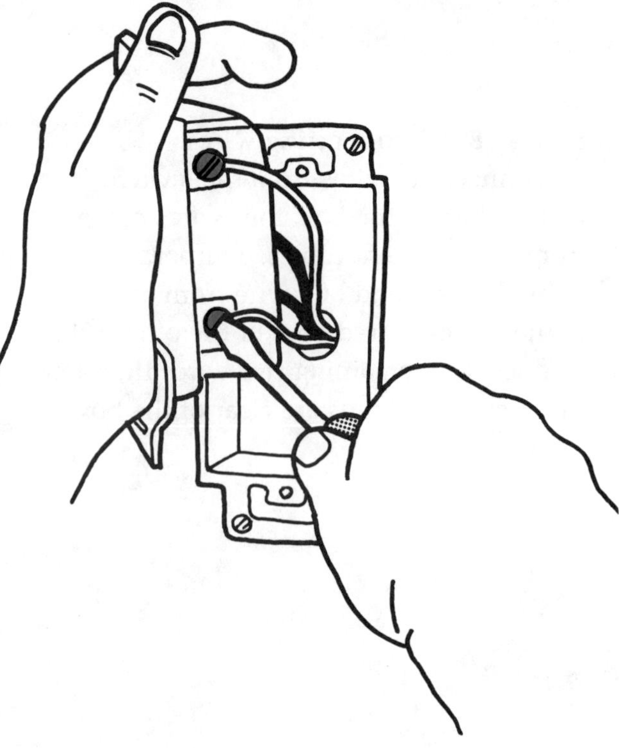

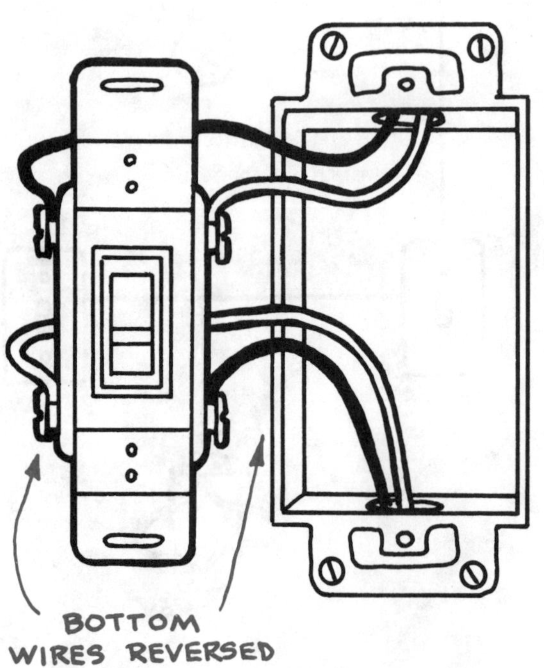


**Step 4-21.**
**Checking the installation.**
If the new four-way switch doesn't work, turn the breaker off, pull the switch from the box, and disconnect the two wires on the bottom of the switch. Interchange the wires and reconnect them to the terminals. The switch should work now.

CHAPTER FIVE

# Replacing Receptacles

Duplex receptacles are the most common outlets found in the home. They seldom fail, but sometimes they are broken or become worn to the point where they don't hold plugs securely. Don't try to repair *duplex receptacles;* replace them. Receptacles can be replaced as easily as switches. And like switches, some are made better than others. Always buy high-quality electrical devices.

## Tools & Materials

- ☐ Screwdriver
- ☐ Needle-nose pliers
- ☐ Voltage tester or volt-ohmmeter
- ☐ New receptacle

**Step 5-1. Removing the cover plate.**
Before starting to work on the receptacle, turn the breaker off. Now remove the cover plate.

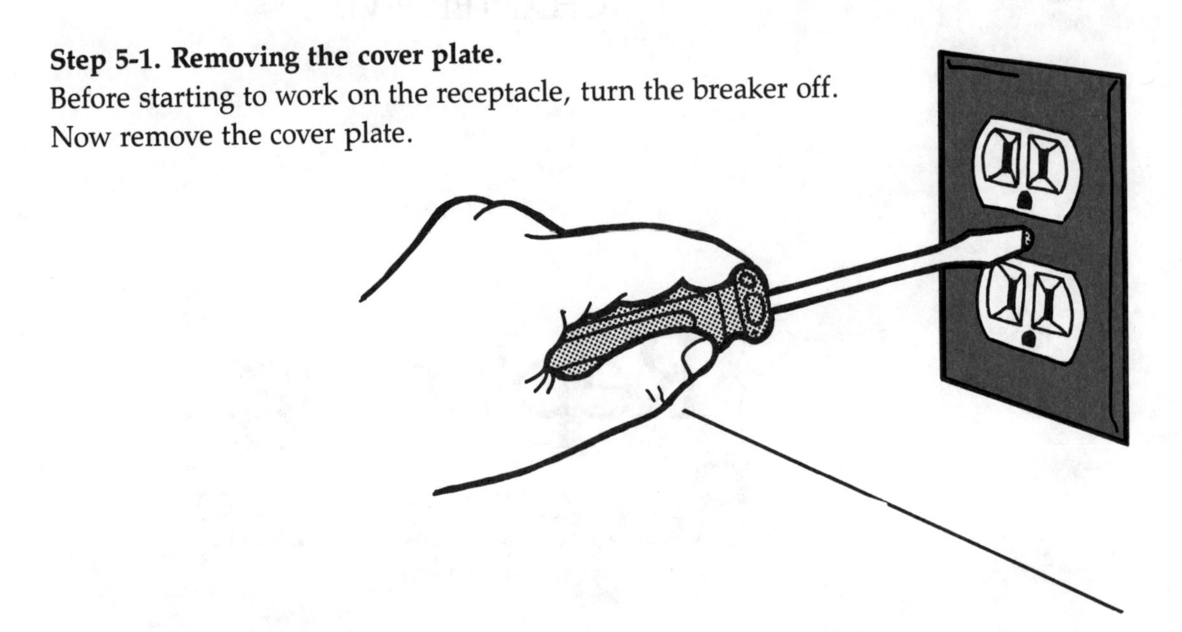

**Step 5-2. Checking for voltage.**
Touch the probes of the voltage tester to the brass and silver terminals. The light on the tester will not light if the circuit is dead.

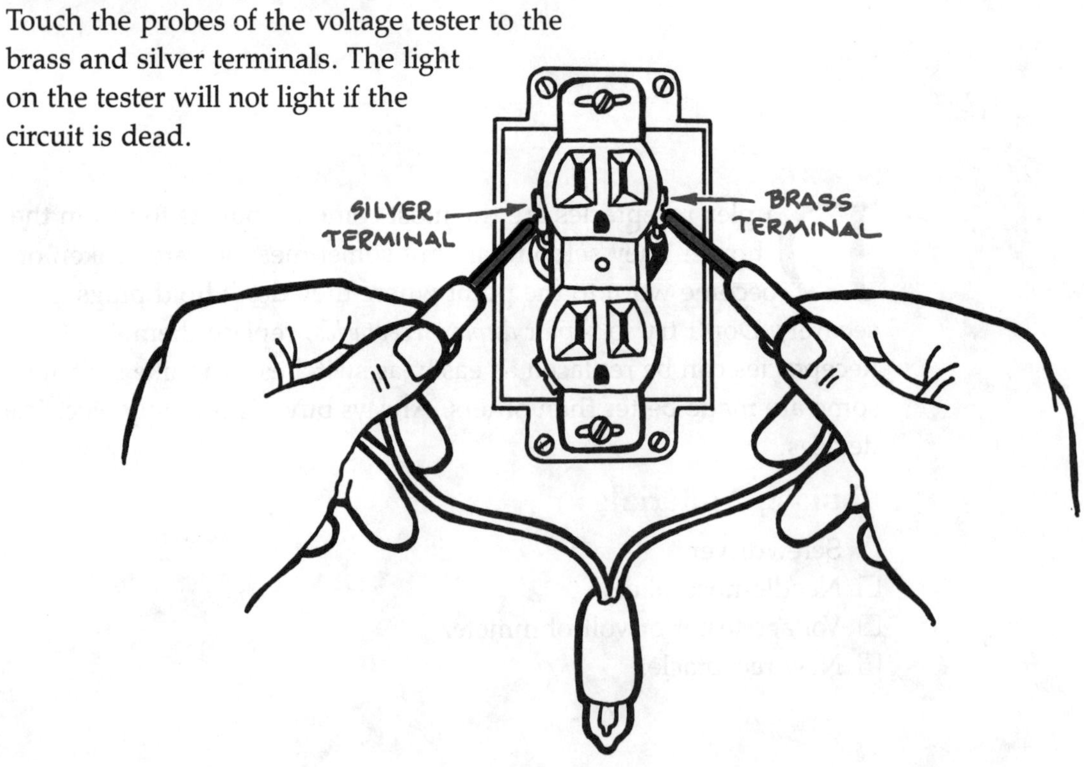

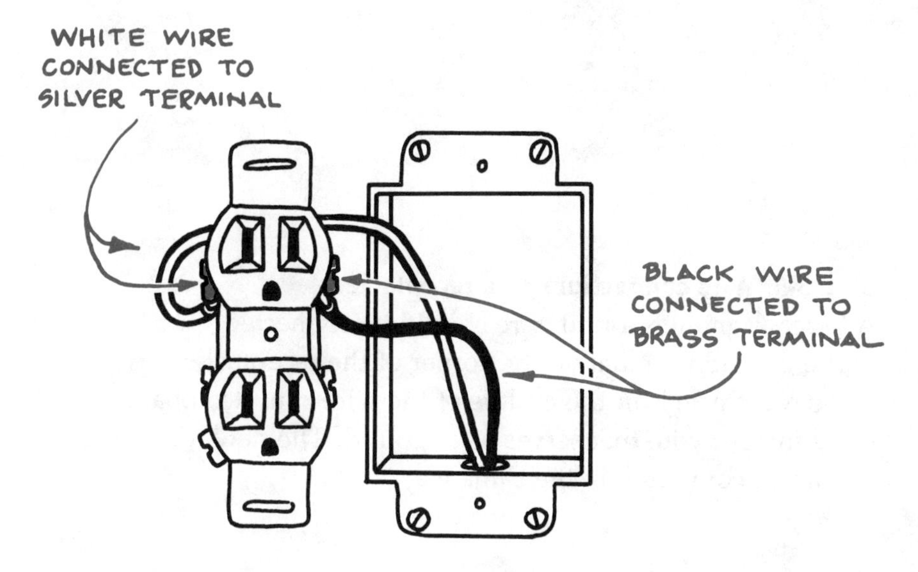


**Step 5-3. Removing a receptacle.**
If the circuit is dead, remove the mounting screws and pull the receptacle from the box. Note that the black wire is connected to the brass terminal and the white wire is connected to the silver terminal.

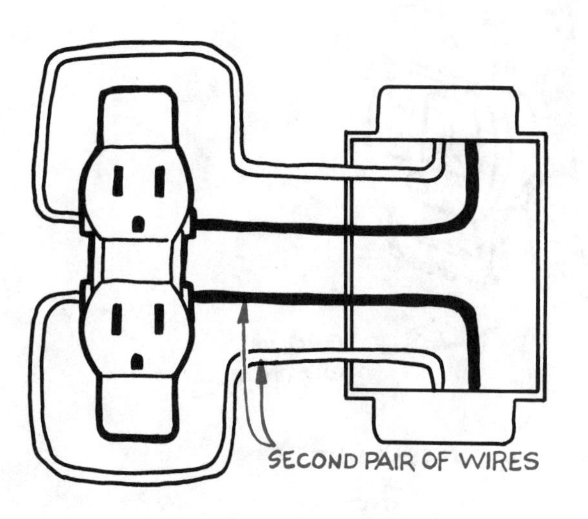


**Step 5-4.**
**Receptacle in middle of the wire run.**
If the receptacle has two black and two white wires, the second pair of wires supplies power to receptacles *downstream* in the circuit.

**Step 5-5. Wire connections to a receptacle.**
A bare, or green, ground wire should be connected to the ground terminal on the corner of the receptacle. If the receptacle is in the middle of the wire run, it probably has a jumper wire that serves as a ground. The new receptacle should be connected in the same way.

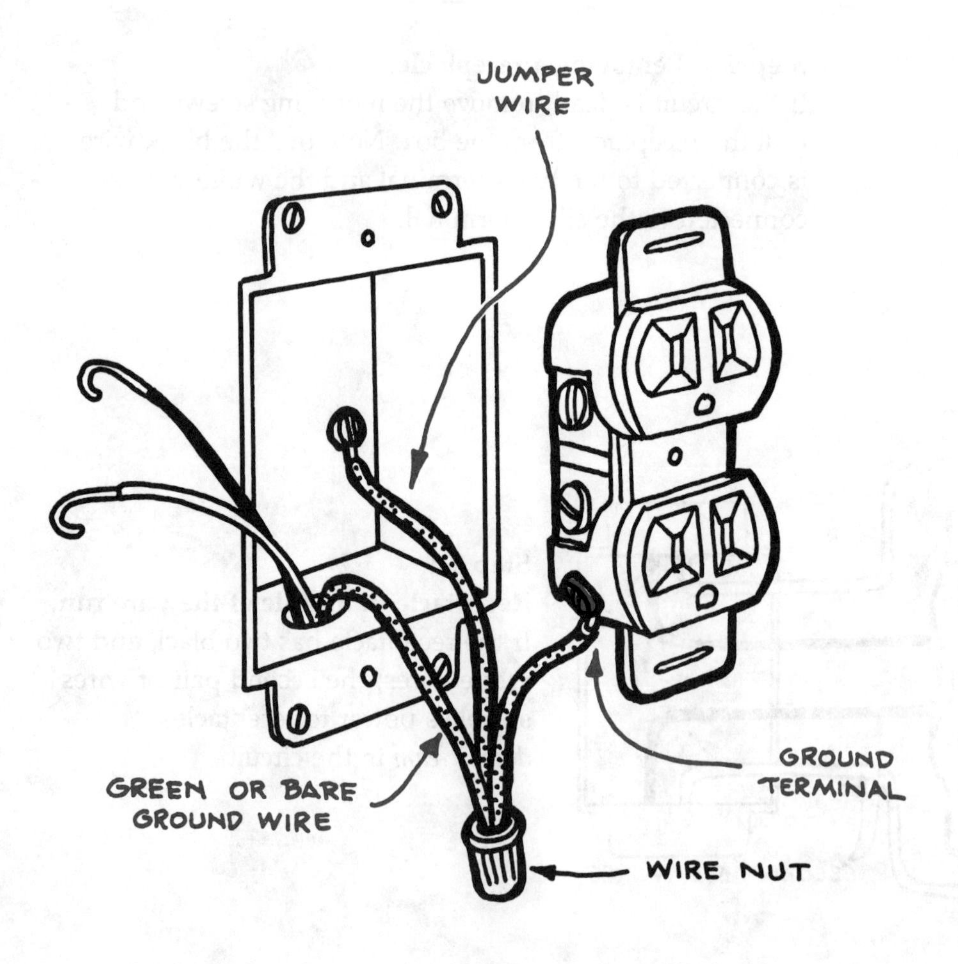

**Step 5-6. Switch-controlled outlet.**
If one end of the receptacle is controlled by a wall switch, the break-off tab between the brass screws has been removed. You need to break the tab on the new receptacle to make it identical to the one you are replacing. Grip the tab—the one on the brass side of the receptacle—with the nose of the pliers and bend the tab back and forth. It should break after a few bends.

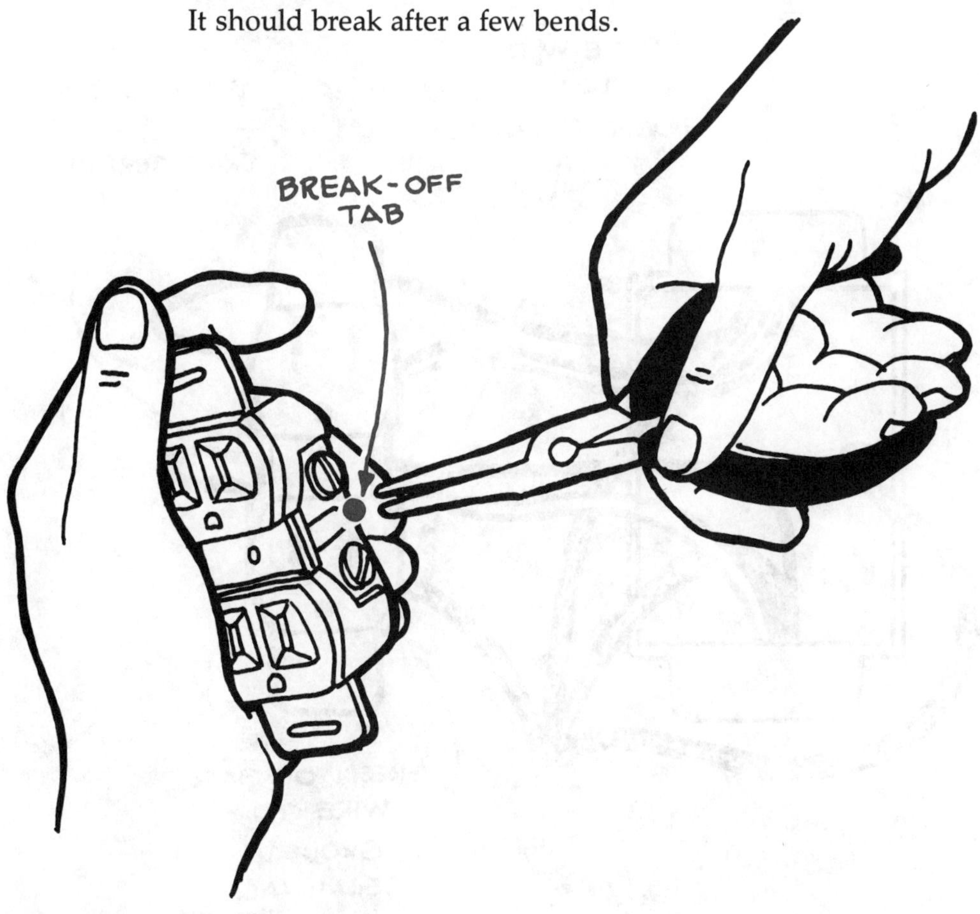

**Step 5-7. Installing a new receptacle.**
Disconnect the wires and attach them to the new receptacle. Connect the black wire to the brass terminal and the white wire to the silver terminal. Connect the ground wire to the ground (usually green) terminal.

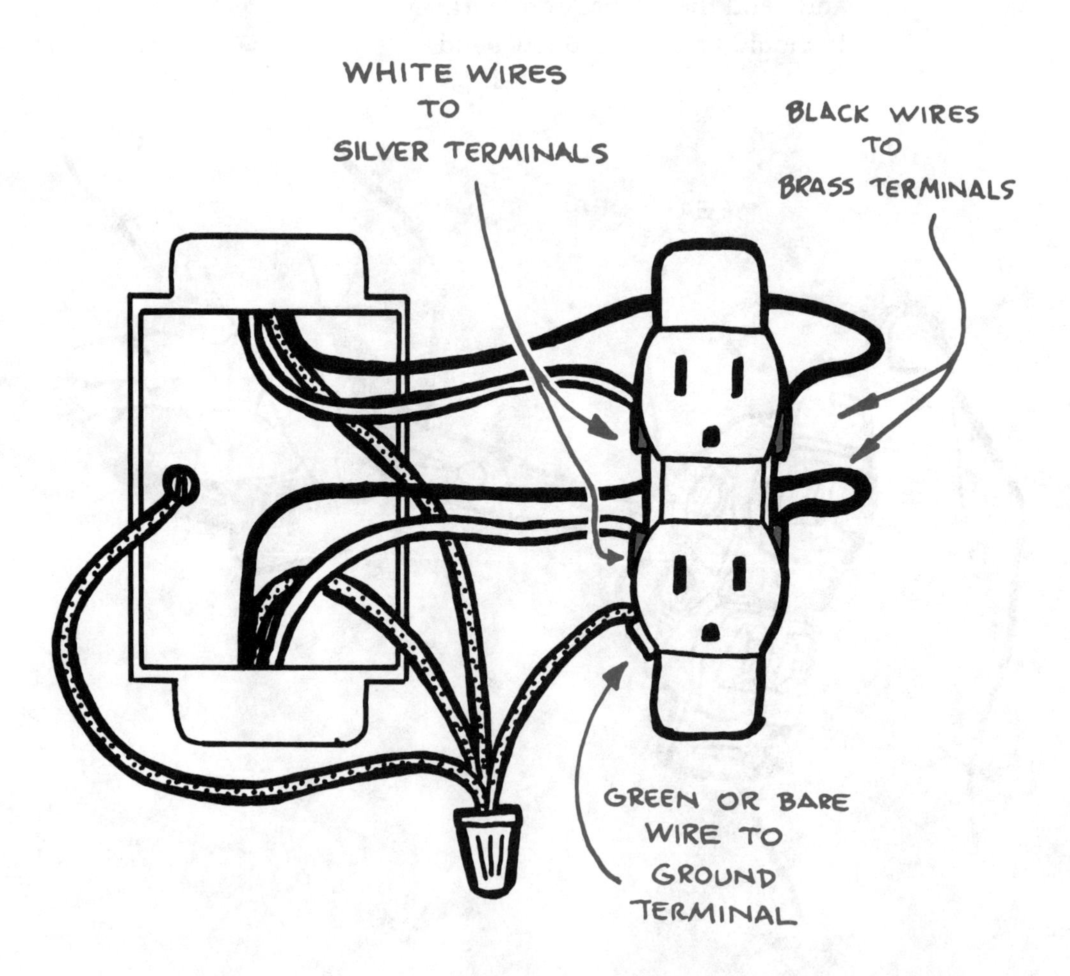

**Step 5-8. Fitting wires into the box.**
Fold the wires as you push the receptacle back in the box.

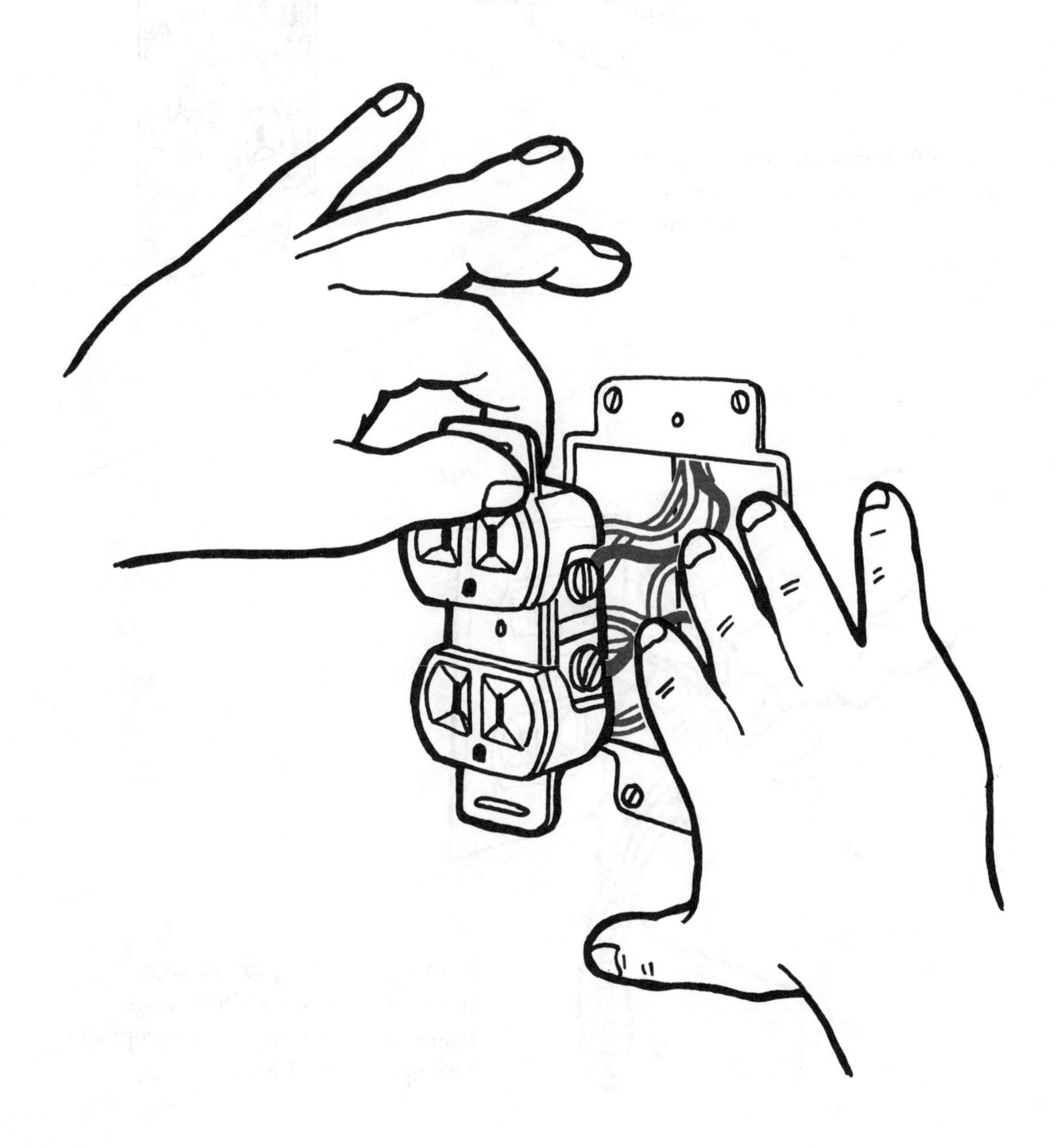

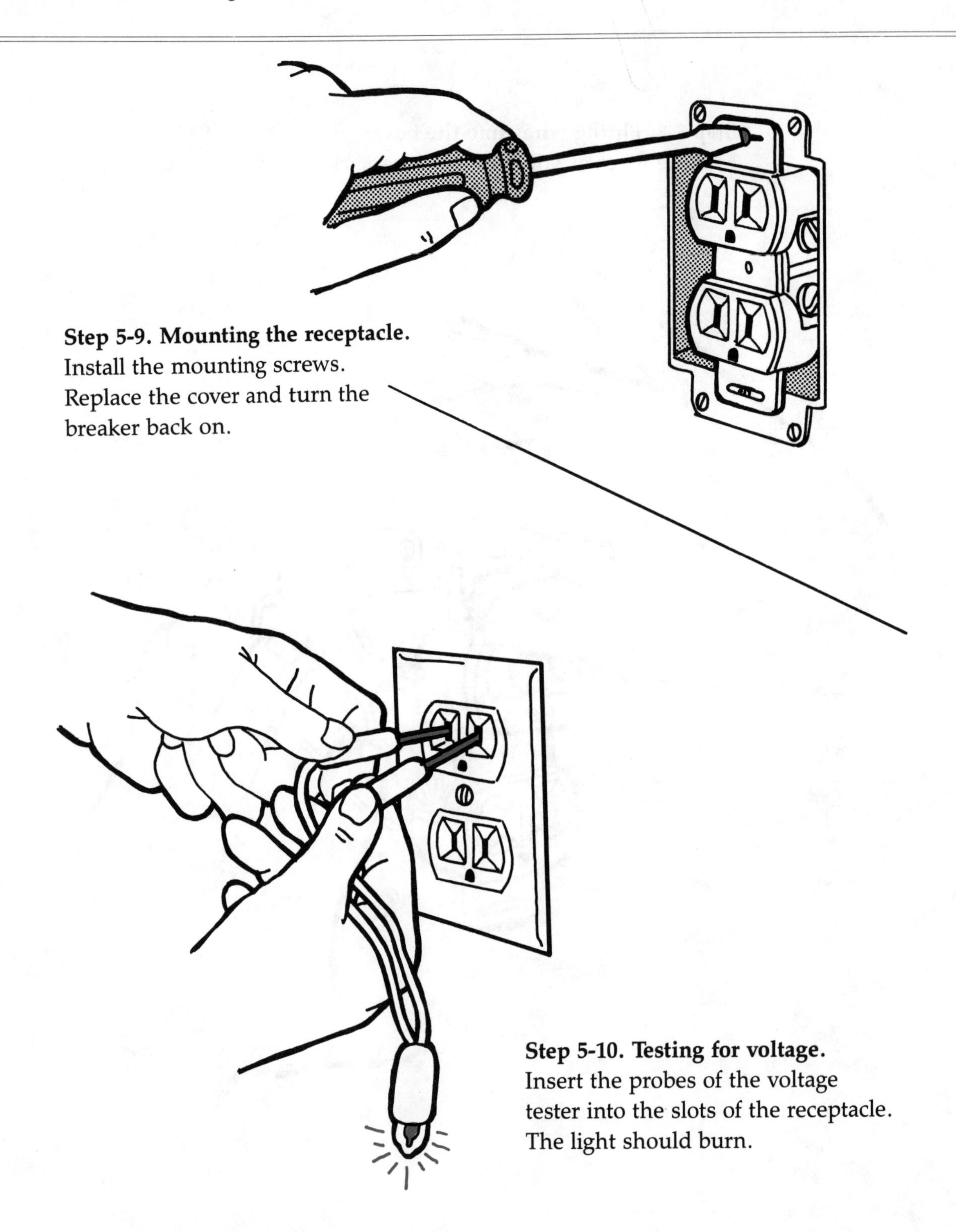

**Step 5-9. Mounting the receptacle.** Install the mounting screws. Replace the cover and turn the breaker back on.

**Step 5-10. Testing for voltage.** Insert the probes of the voltage tester into the slots of the receptacle. The light should burn.

**Step 5-11. Testing ground slot.**
Next, insert one of the probes into the small, arched ground slot, and the other probe into each of the rectangular slots. The light should only burn when the probe is in the smaller of the two slots. This check shows that the receptacle is grounded.

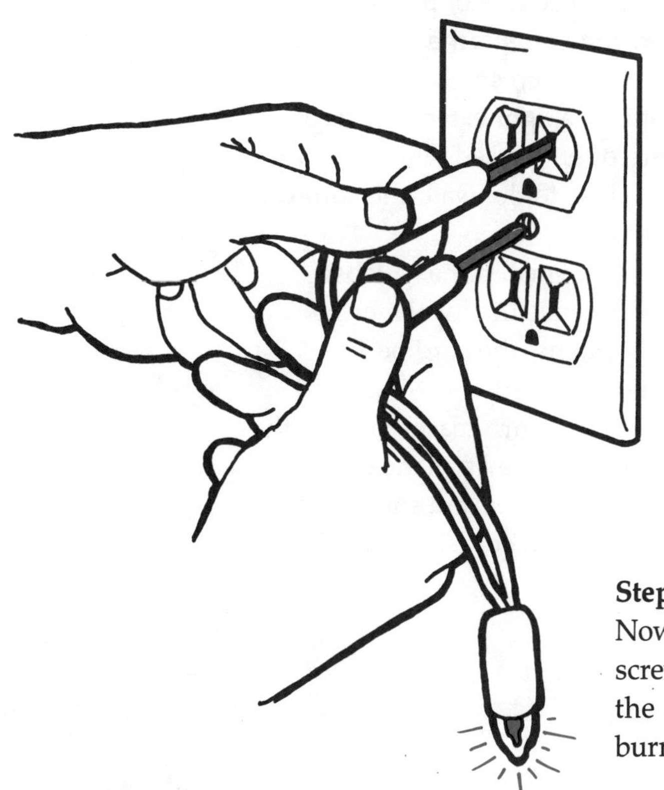

**Step 5-12. Testing the plate for ground.**
Now touch one probe to the cover plate screw and insert the other probe into the slot that had the voltage. If the light burned, the plate is properly grounded.

CHAPTER SIX

# Electrical Cords & Plugs

Electrical cords are the most common source of problems in the home wiring system. Extension cords can be a particular problem because they are sometimes too small for the appliance they are serving. An example is a light extension cord used with an electric heater. Cords are often abused when they are unplugged. If furniture is pushed too close to the wall, the cord can become kinked and damaged at the plug.

A cord should not overheat. If it is too small or has a loose connection, however, heat and time will cause the insulation to become brittle and crack. When this happens, the cord becomes dangerous and should be replaced—not repaired.

Some of the danger signs to look for are burn marks around the plug, loose prongs, intermittent operation of the appliance, and physical damage to the plug or insulation. Repairs are simple and inexpensive. Don't take the risk of using worn cords or faulty plugs.

## Tools & Materials

- ☐ Small screwdriver
- ☐ Wire cutters
- ☐ Wire stripper
- ☐ New plug or cord

**Step 6-1. Removing a plug from an outlet.**
When unplugging a cord, grip the plug between your thumb and fingers and pull it from the receptacle. Pulling on the cord can cause the wires to stretch and break.

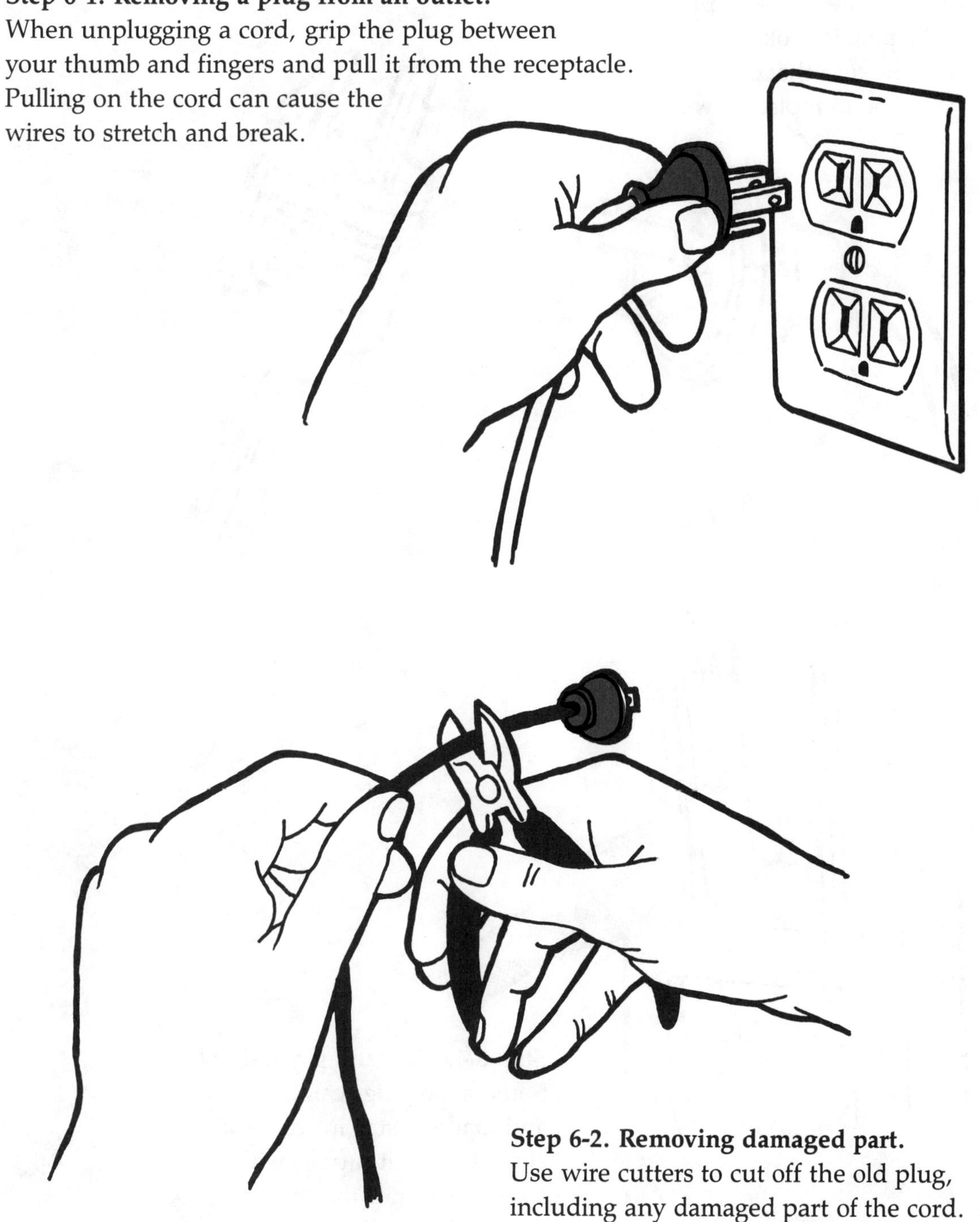

**Step 6-2. Removing damaged part.**
Use wire cutters to cut off the old plug, including any damaged part of the cord.

**Step 6-3. Replacing a plug.**
If the plug is broken, or if it is the molded kind that cannot be taken apart, you need to replace it with a new one.

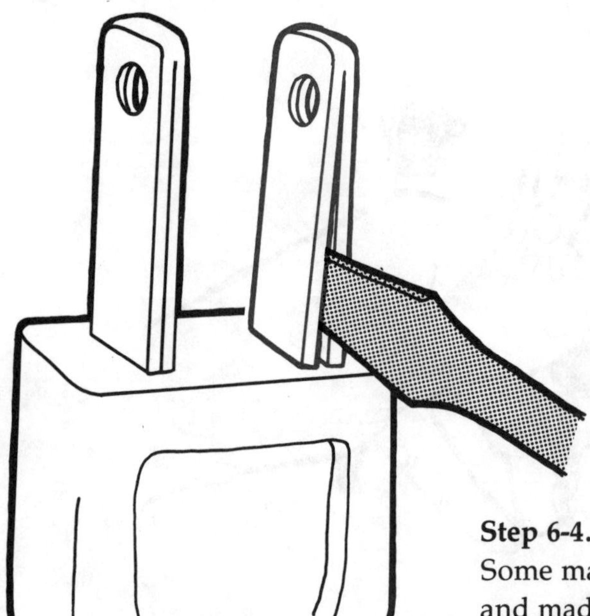

**Step 6-4. Tightening a male plug.**
Some male plugs can be tightened and made to stay in receptacles by spreading split prongs with the tip of a screwdriver. If the prongs are solid, spread them a little farther apart.

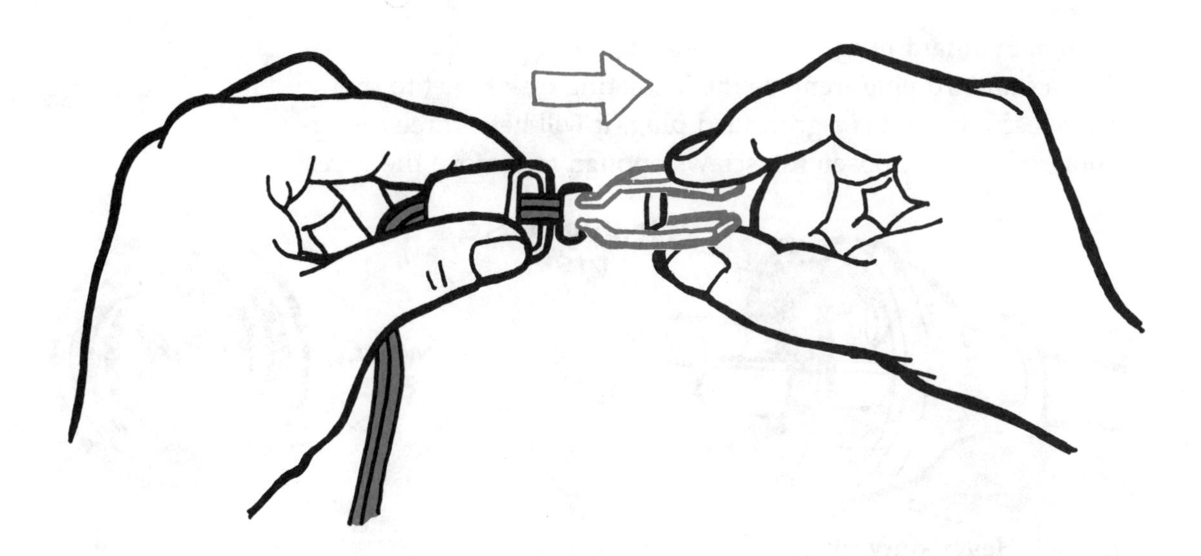

**Step 6-5. Quick-connecting plug.**
If the plug isn't damaged, and can be disassembled, you need to remove the old wires. The way you do this depends on the type of plug you have. If it is a quick-connecting type, press the prongs together and pull the prong assembly and wire from the cover. Now spread the prongs apart to release the wire.

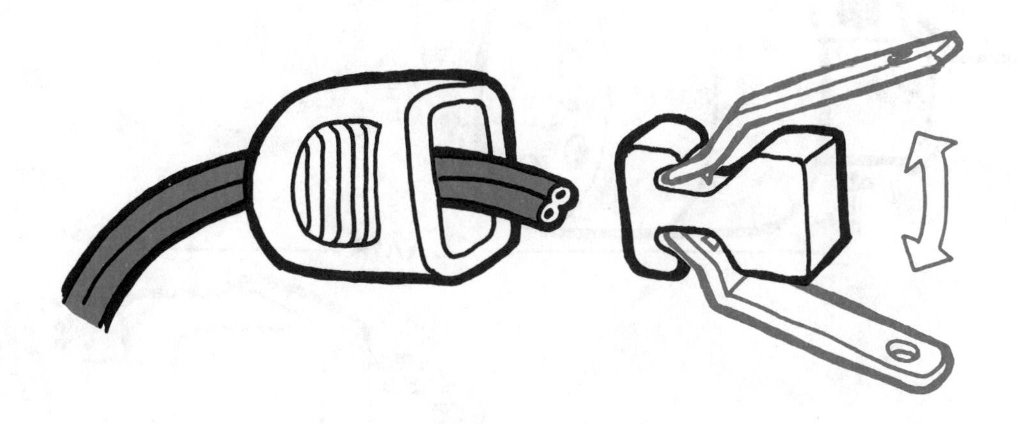

**Step 6-6. Standard plug.**

If it is a standard plug, remove the insulating disk to get to the terminal screws. If it is a grounded plug, it will have three prongs (if not, only two). Loosen the screws enough to remove the wires.

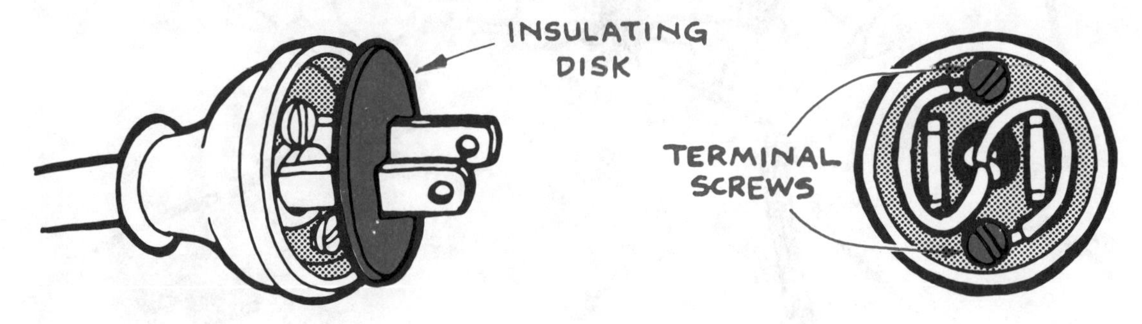


**Step 6-7. Heavy-duty plug.**

If the plug is a heavy-duty type, it probably has some kind of clamp that grips the back part of the plug to the cord. The clamp protects the connections inside the plug if any strain is put on the cord. The front part of the plug is often held in place by three screws. Loosen the clamp screw and the three screws on the end to separate the plug. Then loosen the terminal screws and remove the wires.

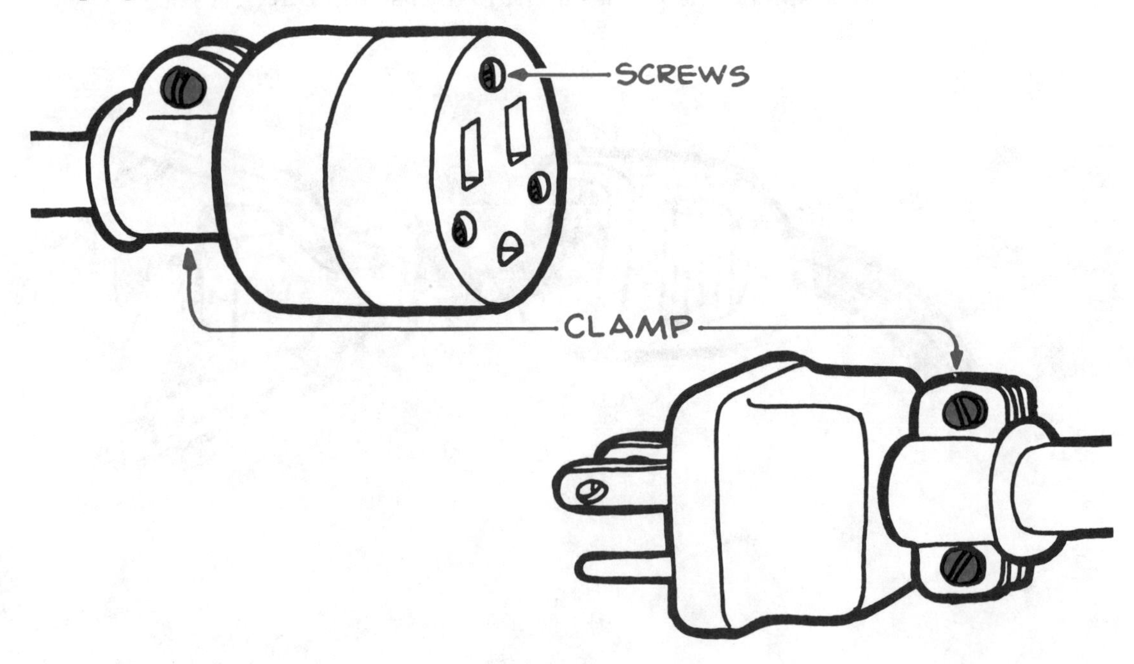

**Step 6-8. Installing a quick-connecting plug.**
To install the quick-connecting plug, cut the end of the cord straight across. No bare wire should protrude past the insulation. Stick the wires through the plug cover and then into the opening in the prong assembly. Press the prongs together to make the electrical connection.
Now push the prong assembly back into the plug cover.

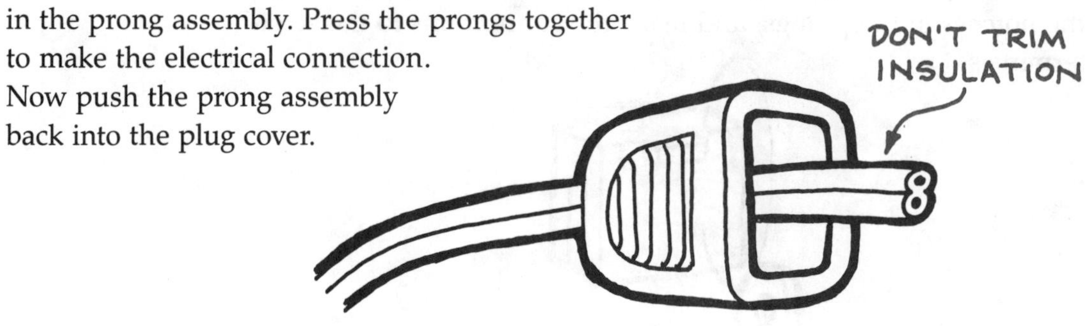


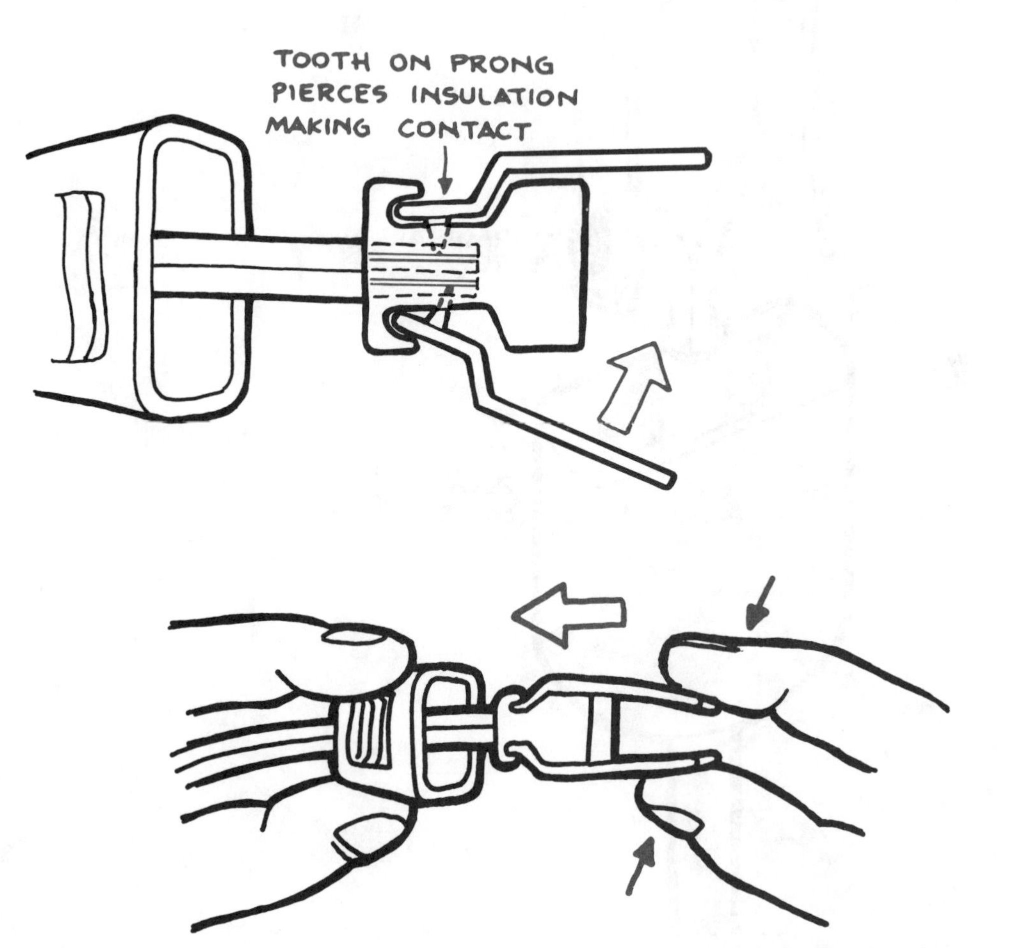

**Step 6-9. Replacing a male plug.**
One type of male plug can be installed by prying out the anchor with a small screwdriver and pulling out the prongs. Thread the cord through the plug cover, place the bare ends of each wire in the notches in the prongs, and make the connections to the screw terminals.

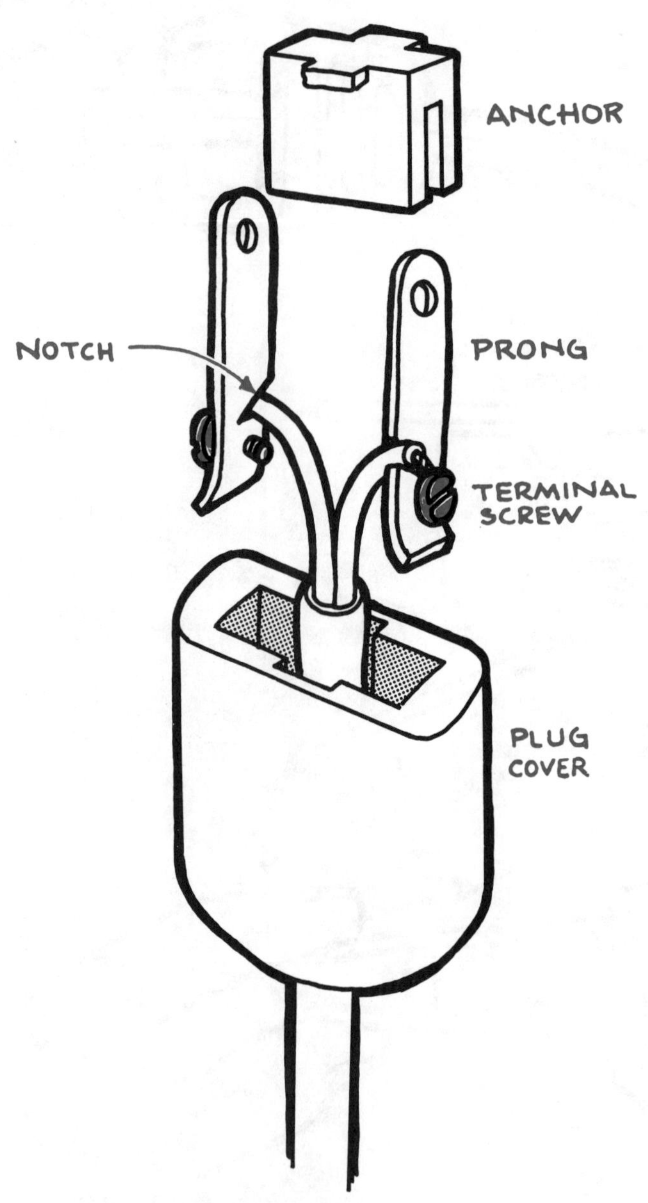

**Step 6-10. Replacing a female plug.**
If the plug has screw terminals, loosen the top screw. Then pry the top from the bottom with a screwdriver. Thread the cord through the plug cover and tie an Underwriters knot (see step 6-14). Insert the bare ends of each wire through the holes in the contacts (metal bars) and connect them to the terminal screws. Reassemble the plug.

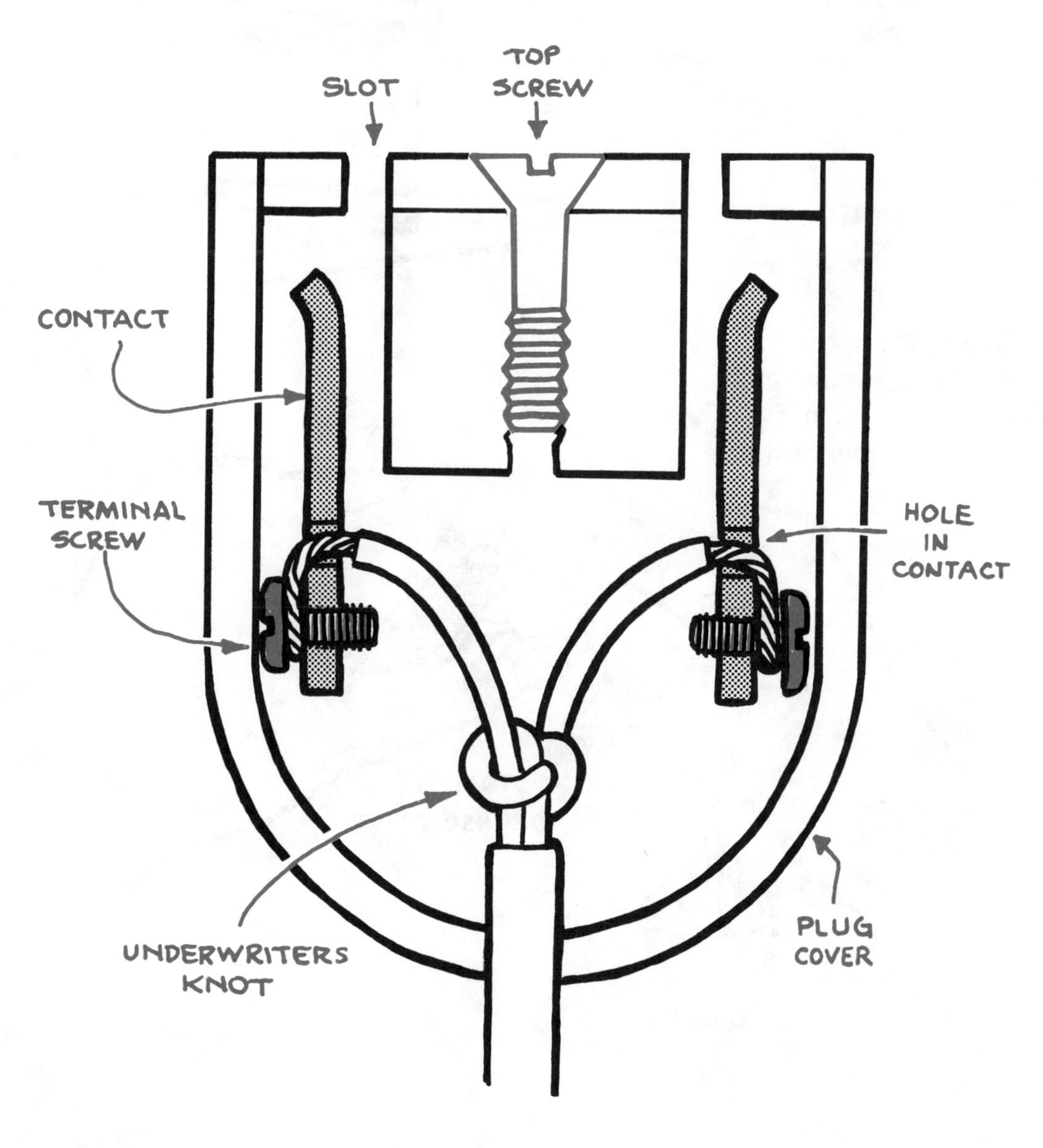

**Step 6-11. Preparing the cord.**
To install a standard plug or heavy-duty plug, insert the end of the cord through the back of the plug. Next, remove about 2 inches of the outer insulation. Lamp cords do not have this cover. Separate lamp cords about 2 inches from the ends.

**Step 6-12.**
**Removing insulation.**
Strip about 1/2 inch of insulation from each wire.

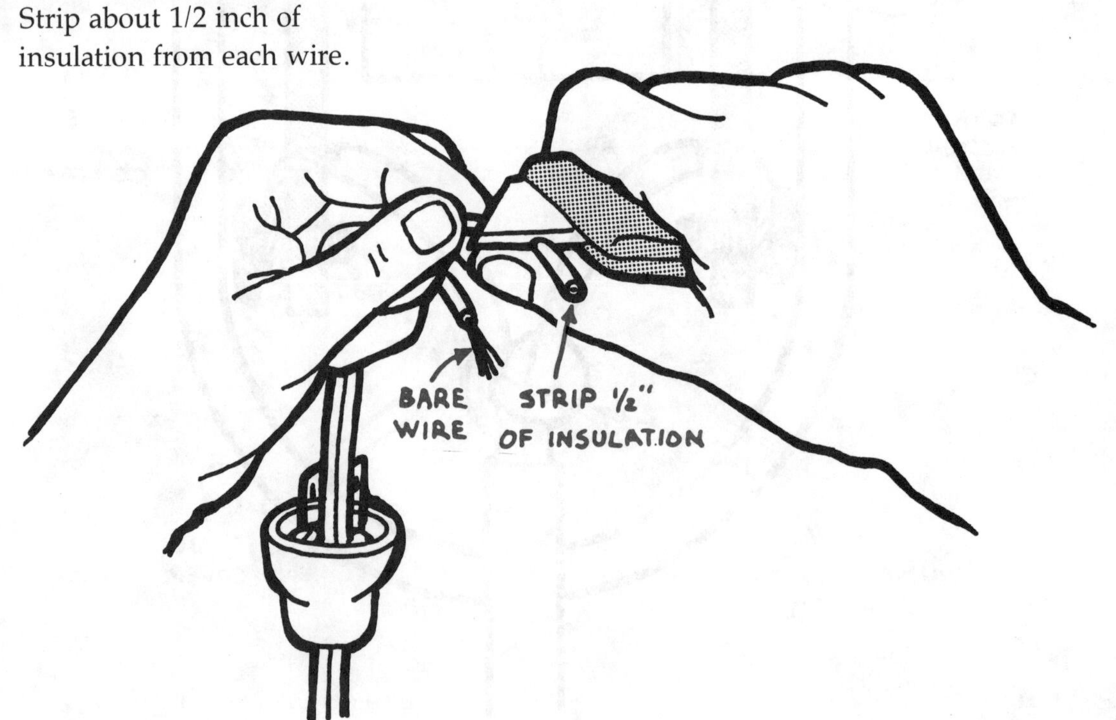

**Step 6-13. Binding strands.**
Twist the strands of each wire so that they bind together. Binding the strands keeps them from spreading out when they are tightened under the screws.

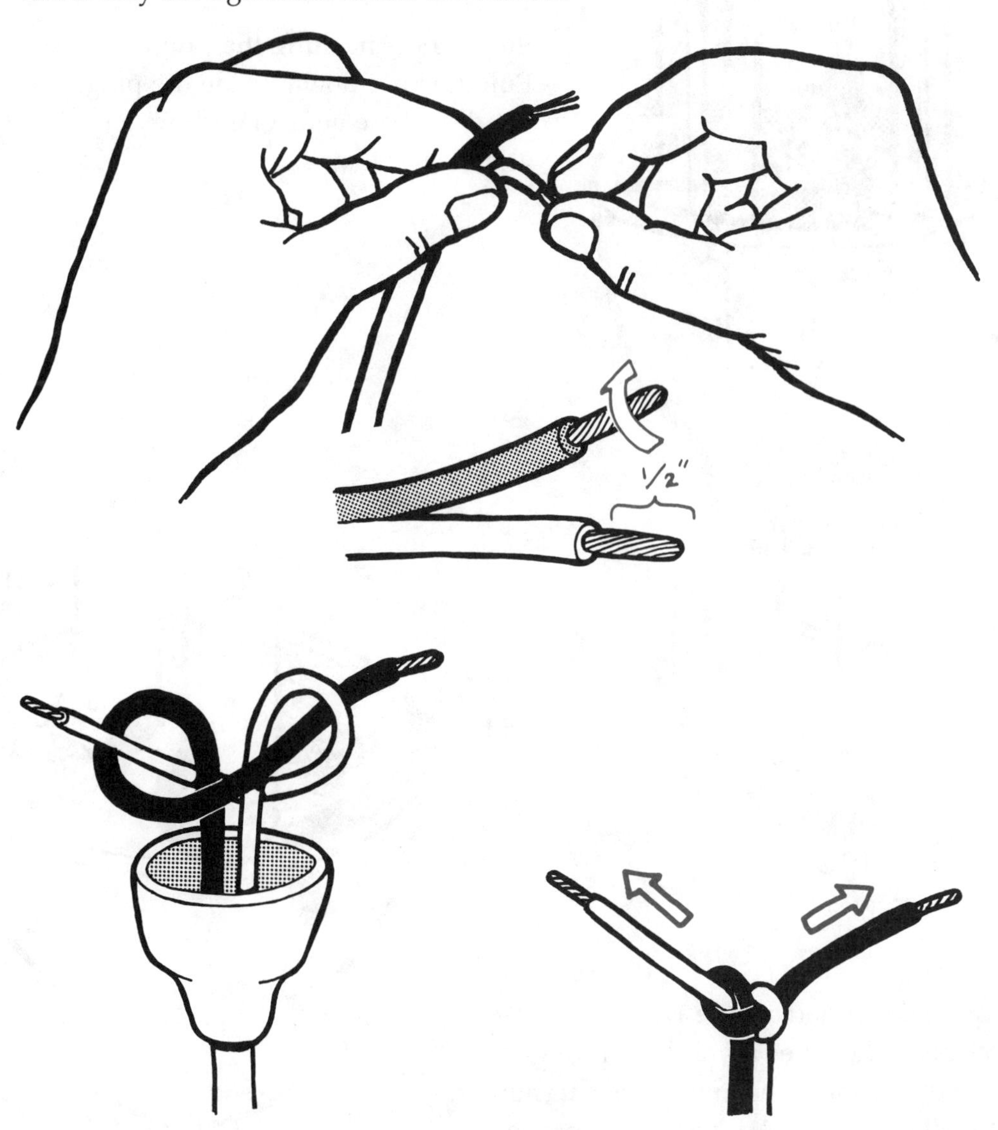


**Step 6-14. Tying an Underwriters knot.**
If the plug has no strain-relief clamp, you should tie an Underwriters knot to take up the strain. Tighten the knot so that the free ends are of equal length.

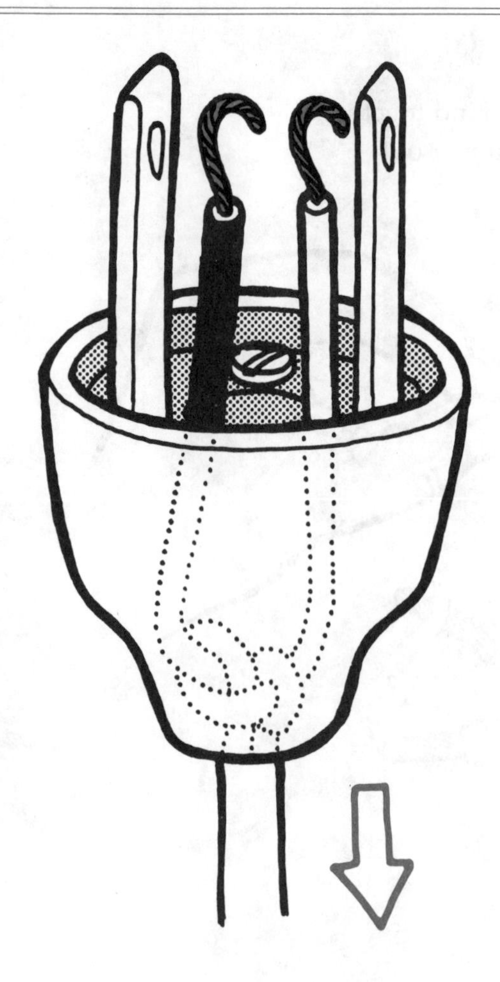

**Step 6-15. Attaching the plug.**
Pull the knot down inside the plug. Bend the bare ends of each wire into a clockwise loop.

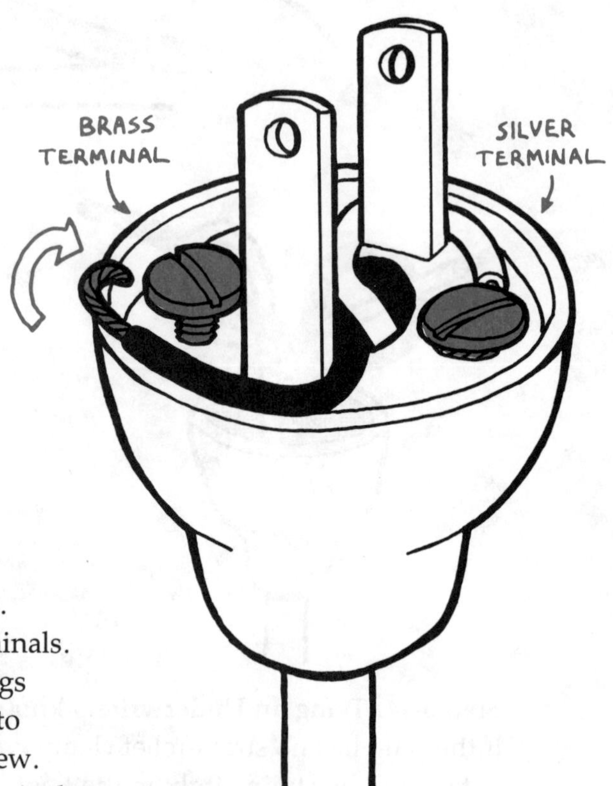


**Step 6-16. Connecting the wires.**
Now bend the wires around the prongs. Make the connections to the screw terminals. If the plug is polarized, one of the prongs will be smaller. Connect the black wire to this terminal. It should have a brass screw. Connect the white wire to the silver terminal.

**Step 6-17. Insulating disk.**
Replace the insulating disk.

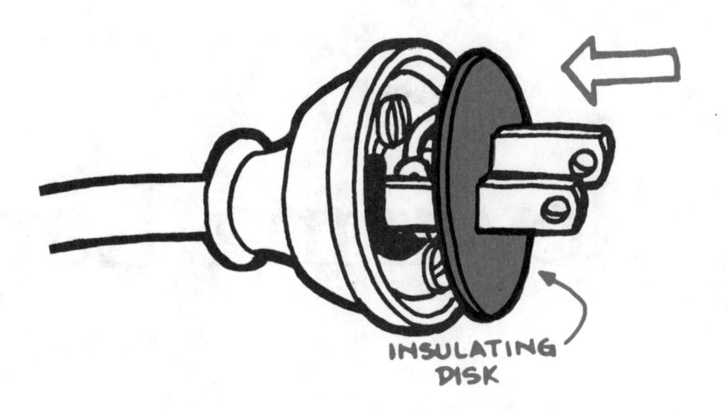


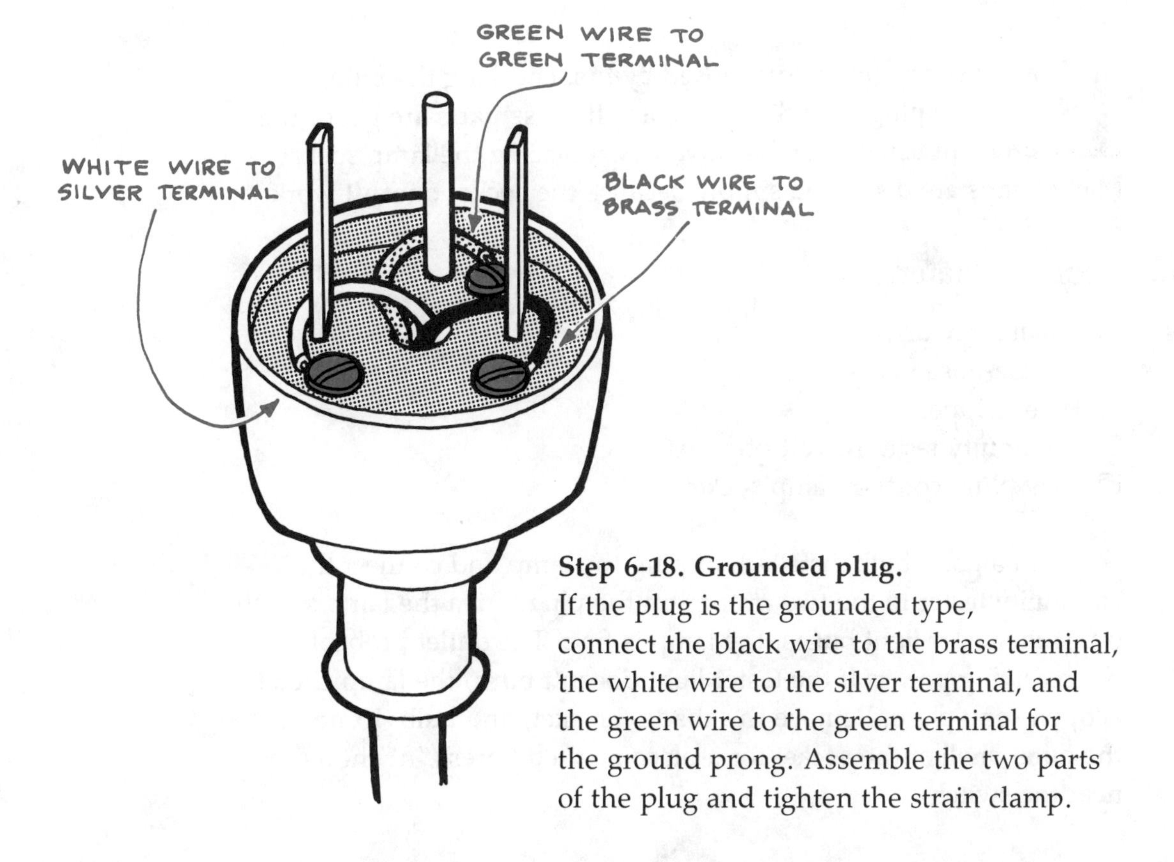


**Step 6-18. Grounded plug.**
If the plug is the grounded type, connect the black wire to the brass terminal, the white wire to the silver terminal, and the green wire to the green terminal for the ground prong. Assemble the two parts of the plug and tighten the strain clamp.

CHAPTER SEVEN

# Rewiring Lamps

Faulty lamps can be diagnosed by first checking the bulb, then the plug, and then the cord. If these parts are good, the problem usually can be solved by replacing the lamp socket. Many lamps are discarded simply because the socket doesn't work.

## Tools & Materials

- ☐ Small screwdriver
- ☐ Needle-nose pliers
- ☐ Wire stripper
- ☐ Continuity tester or volt-ohmmeter
- ☐ New plug, cord, or lamp socket

With a good bulb installed, unplug the lamp and connect the continuity tester to the two prongs on the plug. Turn the lamp switch on. If the tester light burns, the lamp is fine. The outlet probably is not receiving power. If the tester light doesn't burn, the lamp circuit is open someplace. Remove the shade, bracket, and bulb. Examine the outer shell of the socket for a place marked "press." It should be near the switch.

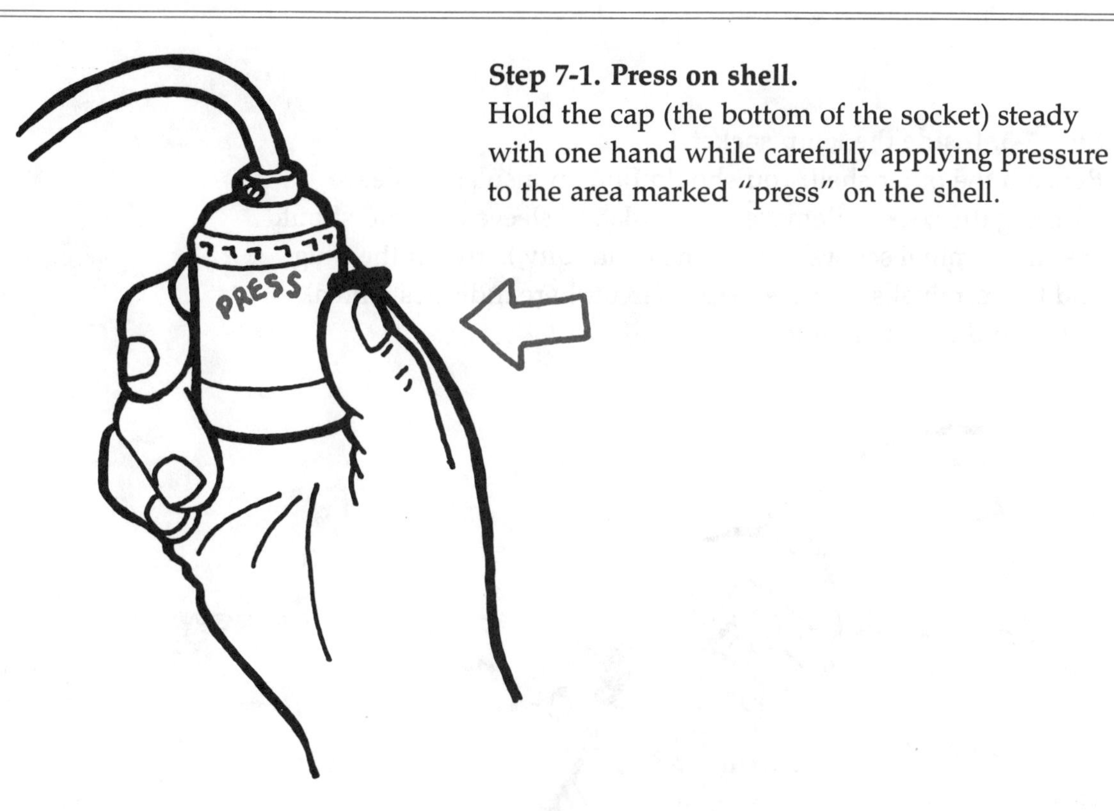


**Step 7-1. Press on shell.**
Hold the cap (the bottom of the socket) steady with one hand while carefully applying pressure to the area marked "press" on the shell.

**Step 7-2. Removing shell.**
Twist slightly and slip the shell from the cap.

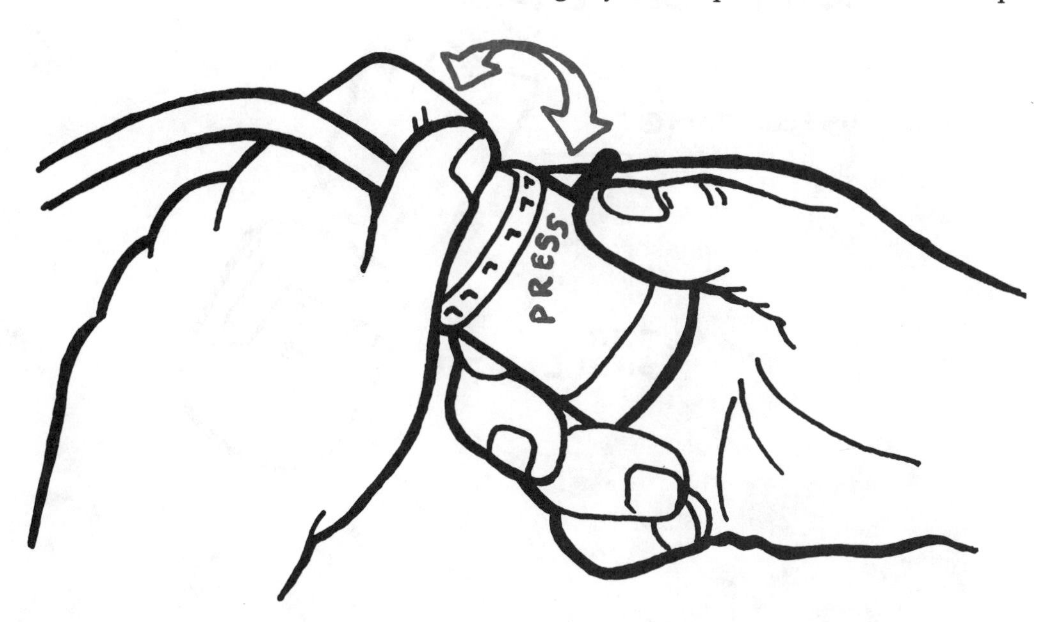

**Step 7-3. Inside the lamp socket.**
Beneath the outer shell, you should find an insulating sleeve covering the socket. Remove the insulating sleeve and you should see the terminal screws. Check the continuity between the plug and the terminal screws. An open circuit here indicates a break in the cord or a bad plug.

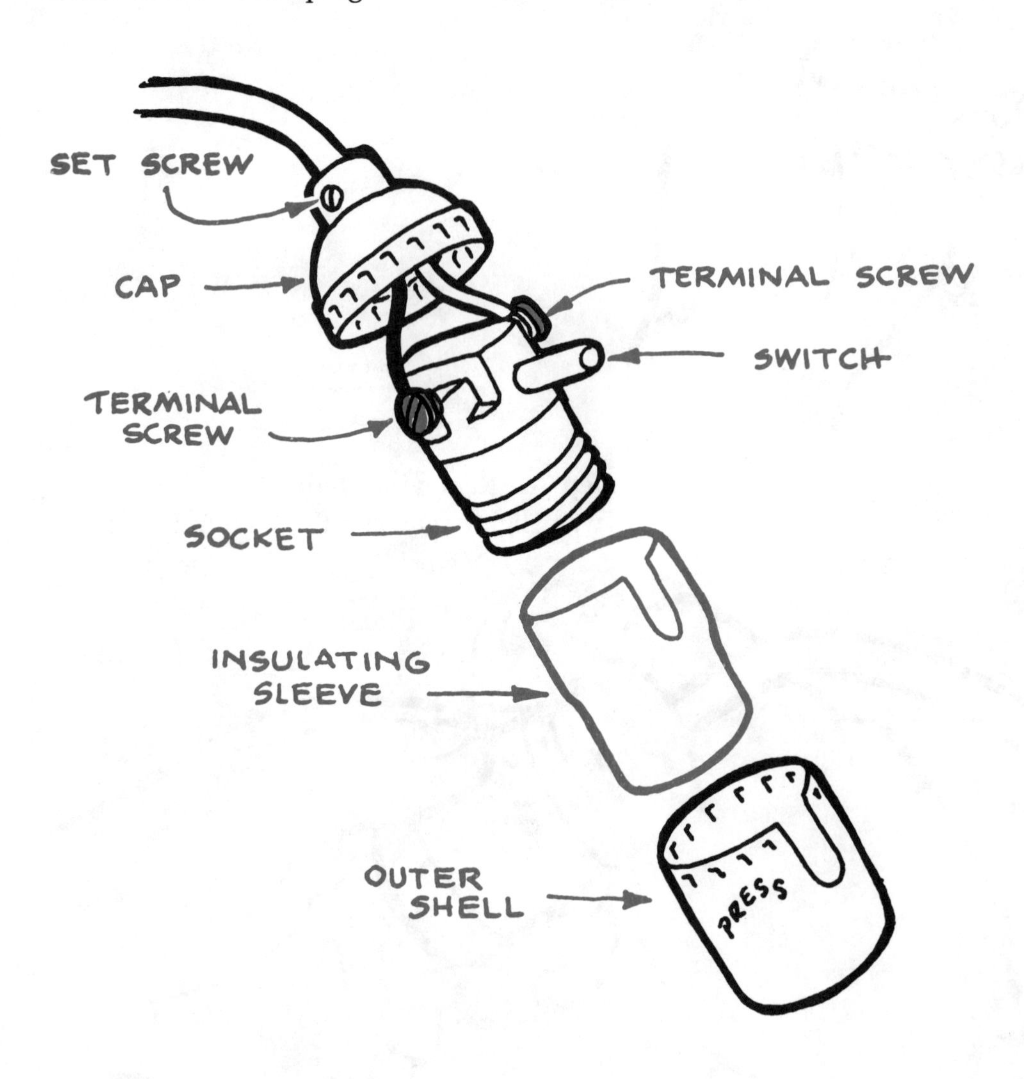

**Step 7-4. Removing the old cord.**
To replace the cord, disconnect the wires from the terminals and remove the socket. Untie the Underwriters knot and pull the old cord out from the bottom of the lamp.

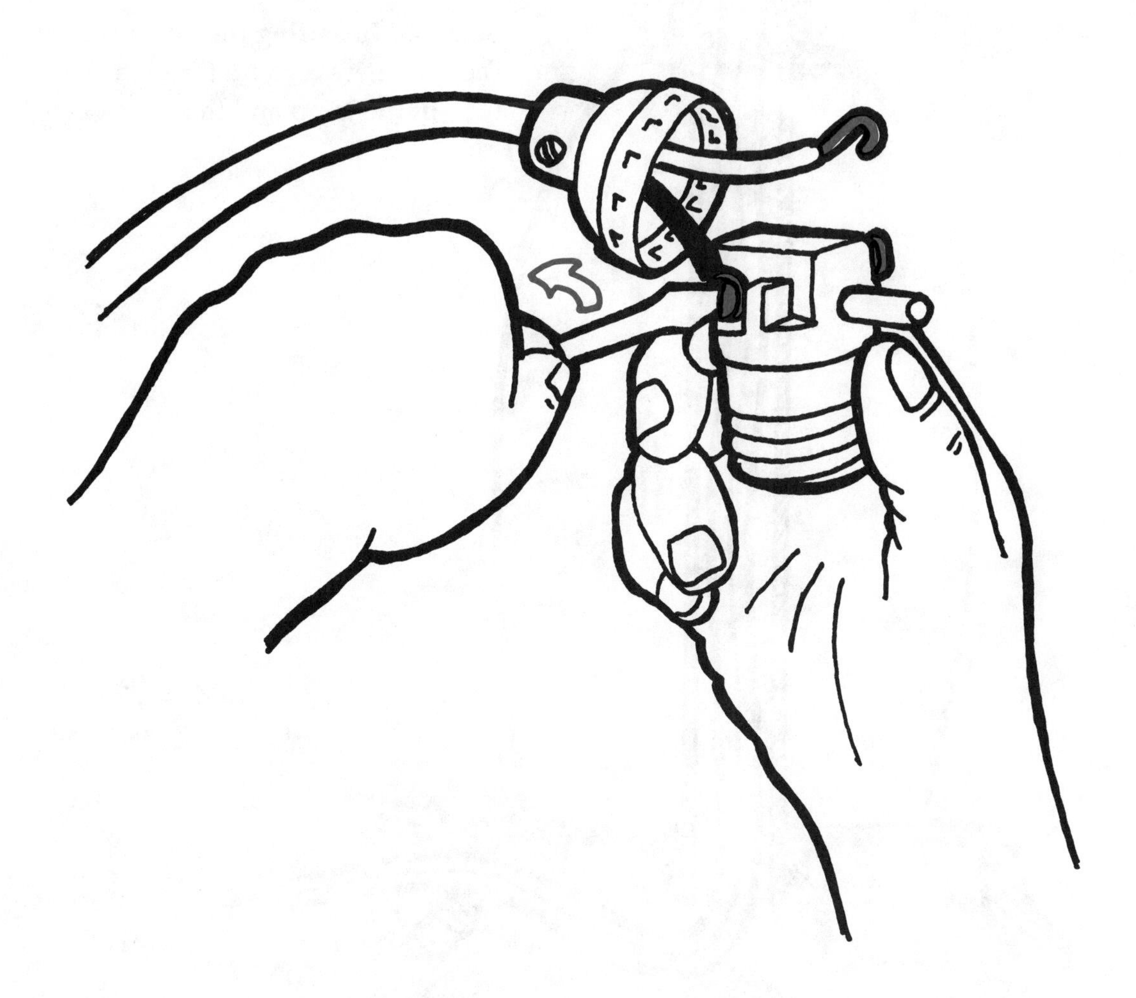

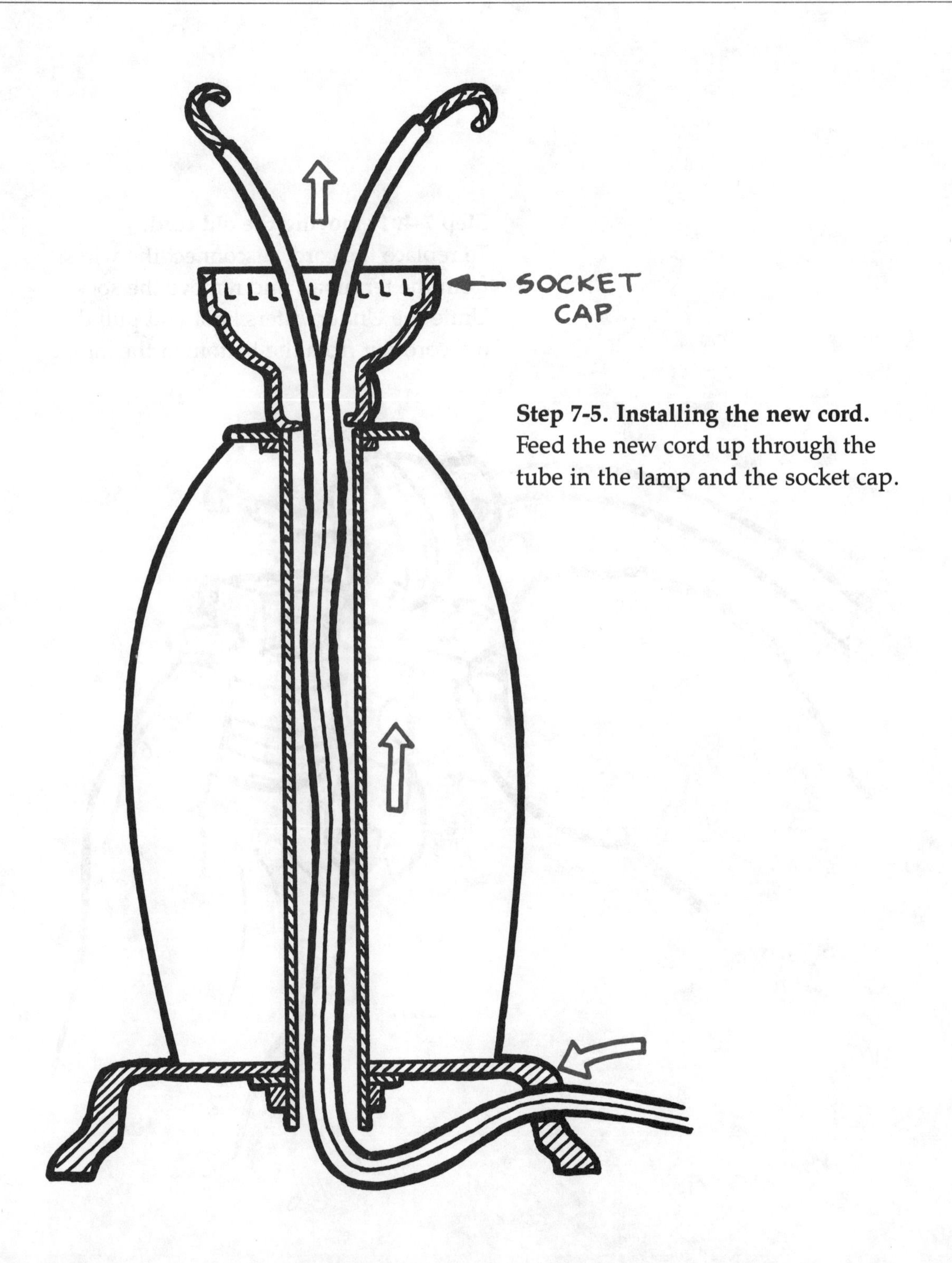


**Step 7-5. Installing the new cord.** Feed the new cord up through the tube in the lamp and the socket cap.

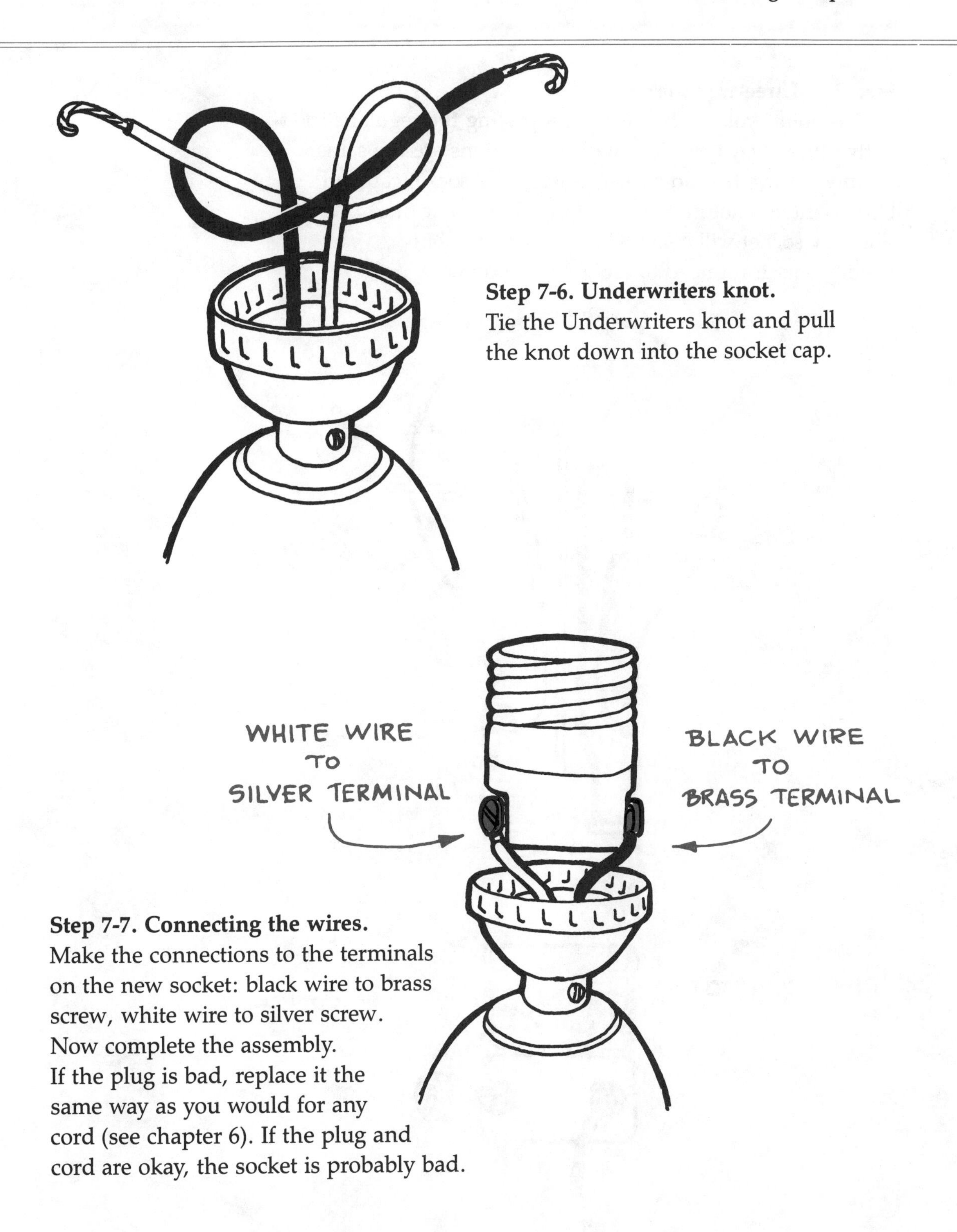


**Step 7-6. Underwriters knot.**
Tie the Underwriters knot and pull the knot down into the socket cap.

**Step 7-7. Connecting the wires.**
Make the connections to the terminals on the new socket: black wire to brass screw, white wire to silver screw. Now complete the assembly.
If the plug is bad, replace it the same way as you would for any cord (see chapter 6). If the plug and cord are okay, the socket is probably bad.

**Step 7-8. Three-way socket.**

At this point, you might consider replacing the regular socket with a three-way type. The wire connections are the same. Simply remove the old socket, leaving the socket cap and Underwriters knot in place, and install the new one. The new socket will come with a socket cap, but usually you don't need to replace the old one.

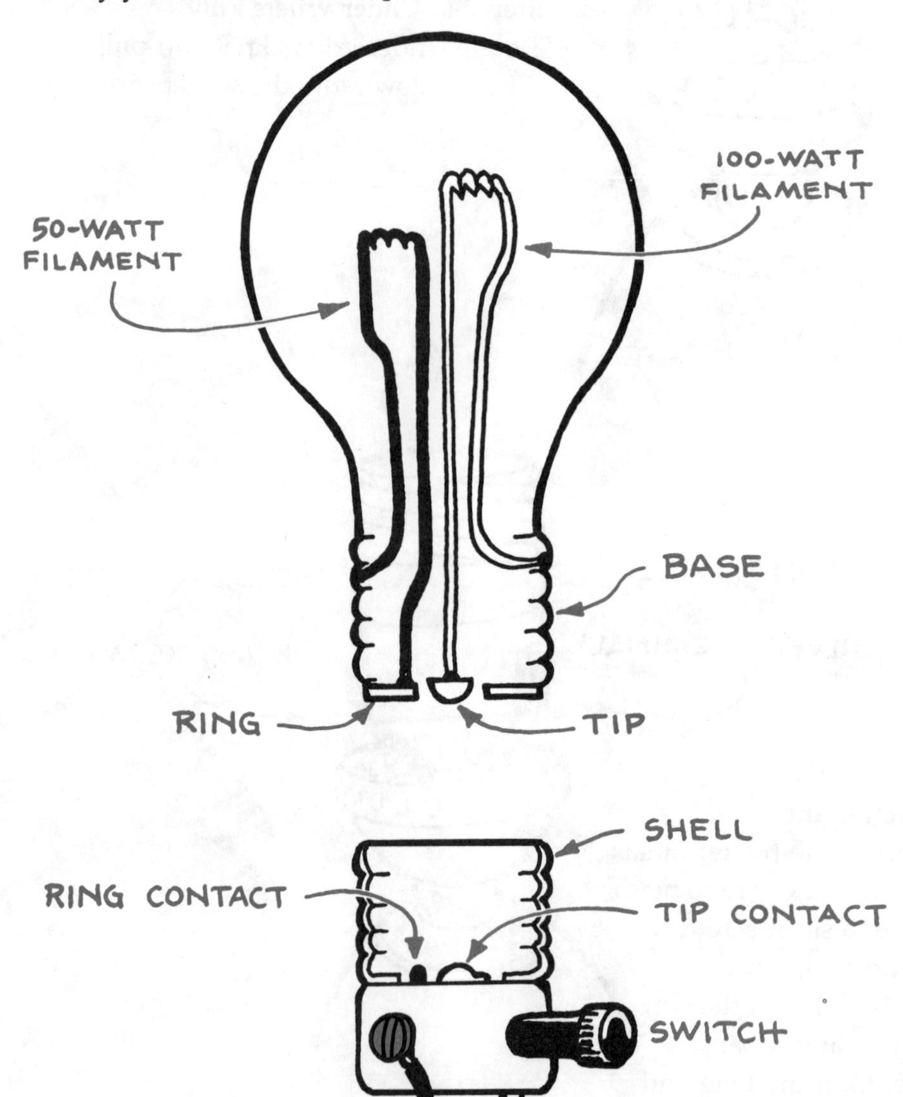

CHAPTER EIGHT

# Replacing Light Fixtures

Light fixtures come already wired, with one black wire and one white wire. You simply connect the black fixture wire to the black hot wire and the white fixture wire to the white neutral wire in the house wiring.

Two types of light fixtures are found in the home. The most common is the incandescent type. This kind uses common light bulbs. The other type is the fluorescent tube light. These fixtures often are recessed in ceilings. Fluorescent lights cost less to operate than regular (incandescent) bulbs, and do not generate as much heat.

## Tools & Materials

- ☐ Screwdriver
- ☐ Needle-nose pliers
- ☐ Wire stripper
- ☐ Wire nuts
- ☐ New fixture, starter, or ballast

**Step 8-1. Removing the old fixture.**
To remove the old fixture, first turn the breaker off. Wall-mounted fixtures usually are held in place by a single, decorative cap nut. Remove the globe and light bulb. Then remove the cap nut and the fixture.

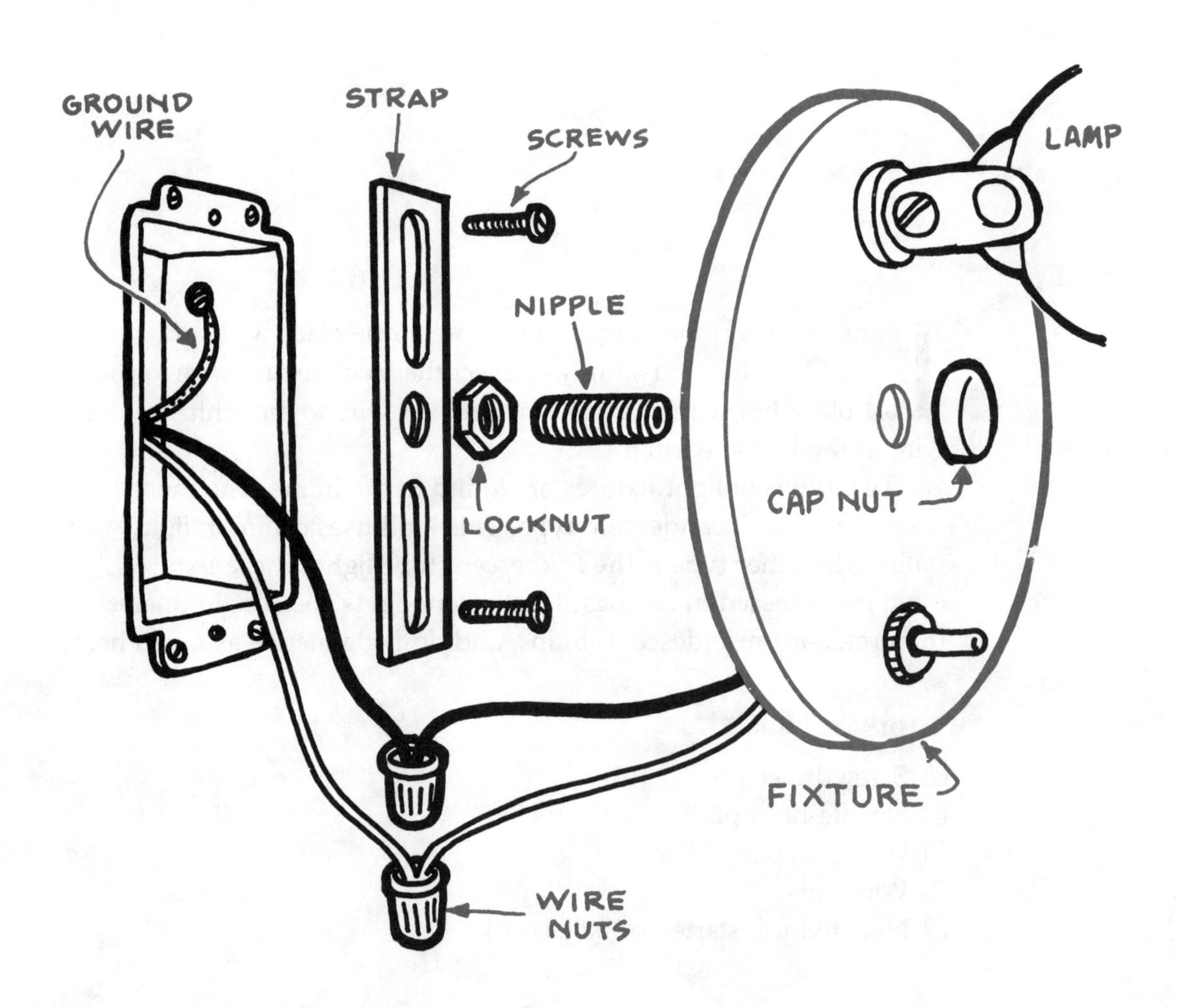

**Step 8-2. Connecting the wires.**
To install the new fixture, strip about 3/4 inch of insulation from each new wire. They might already have been stripped. Use wire nuts to connect the wires: black wire to black wire, white wire to white wire.

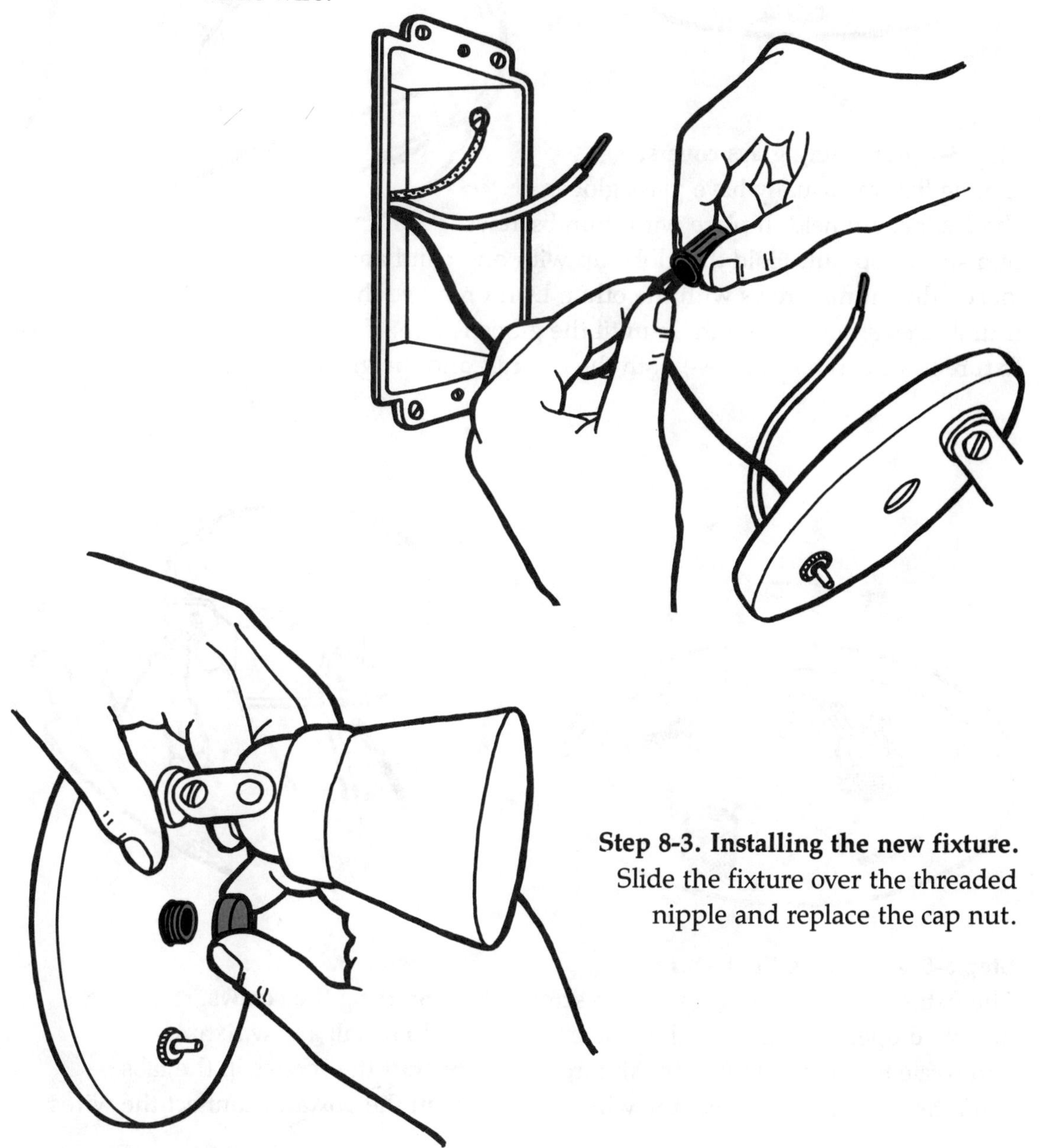

**Step 8-3. Installing the new fixture.** Slide the fixture over the threaded nipple and replace the cap nut.

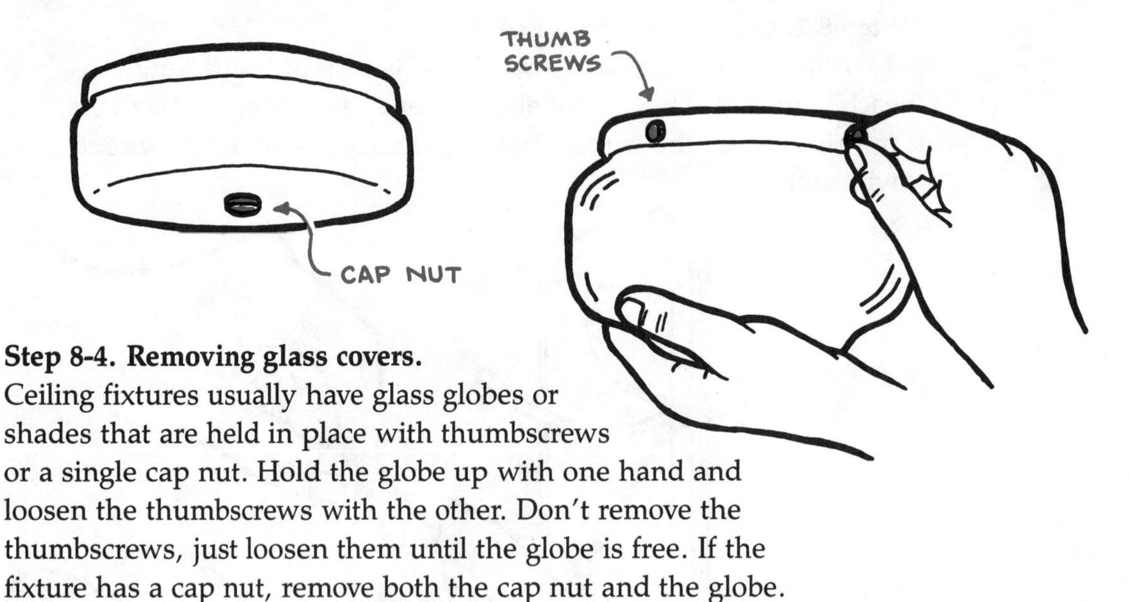


**Step 8-4. Removing glass covers.**
Ceiling fixtures usually have glass globes or shades that are held in place with thumbscrews or a single cap nut. Hold the globe up with one hand and loosen the thumbscrews with the other. Don't remove the thumbscrews, just loosen them until the globe is free. If the fixture has a cap nut, remove both the cap nut and the globe.

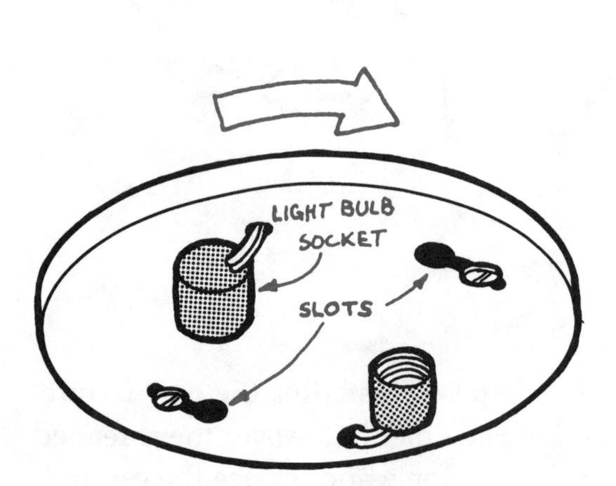


**Step 8-5. Removing the fixture.**
The fixture might be held up by two screws in slotted openings. Loosen these screws and rotate the fixture to free the fixture from the ceiling. Disconnect the wires.

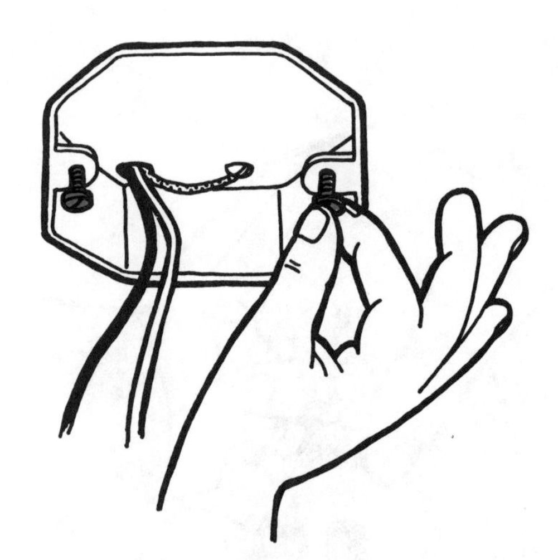

**Step 8-6.**
**Starting the screws.**
To install a new fixture, start the screws in the tabs in the box and connect the wires.

**Step 8-7. Attaching the new fixture.** Place the fixture over the box with the screw heads extending through the slots. Rotate the fixture.

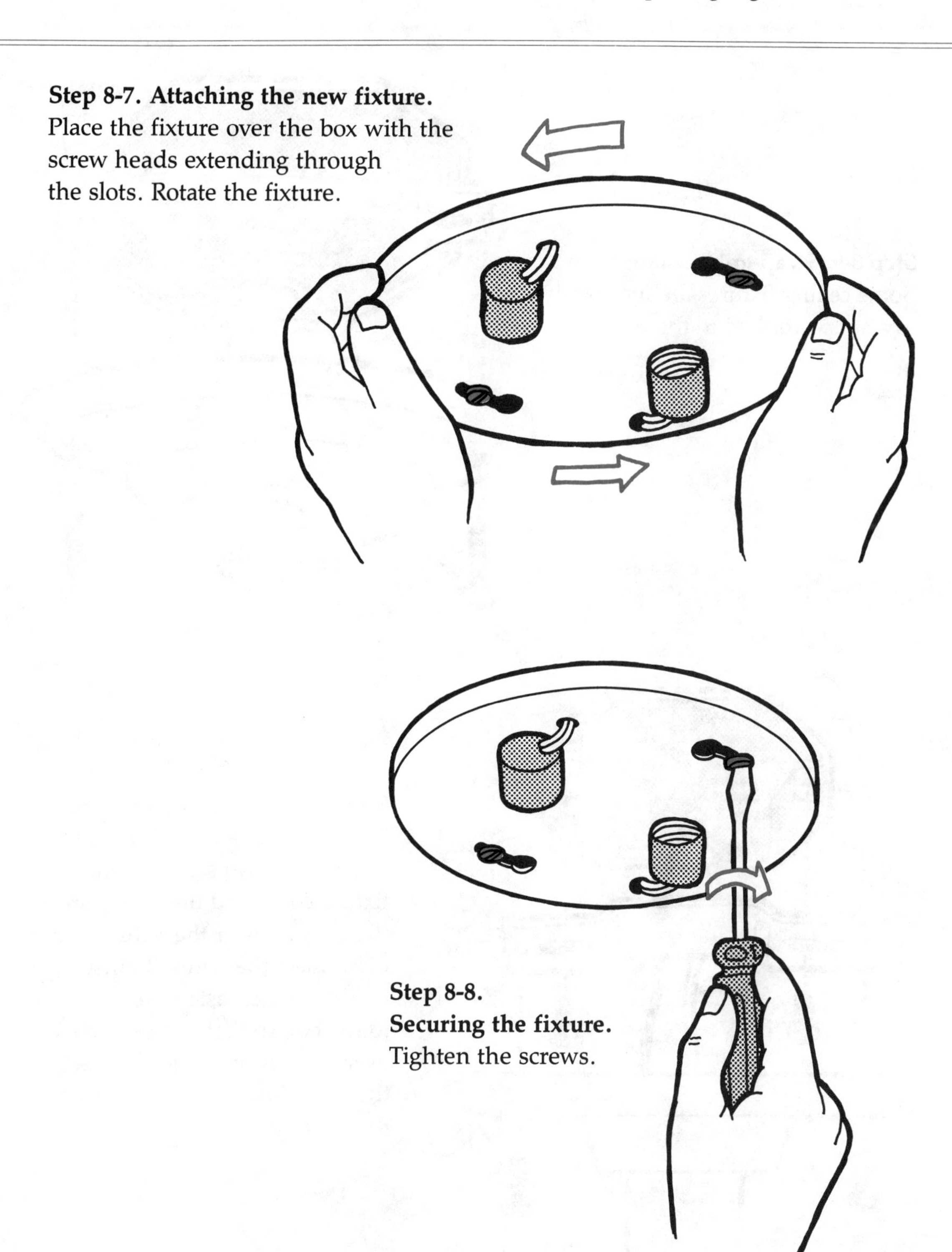

**Step 8-8. Securing the fixture.** Tighten the screws.

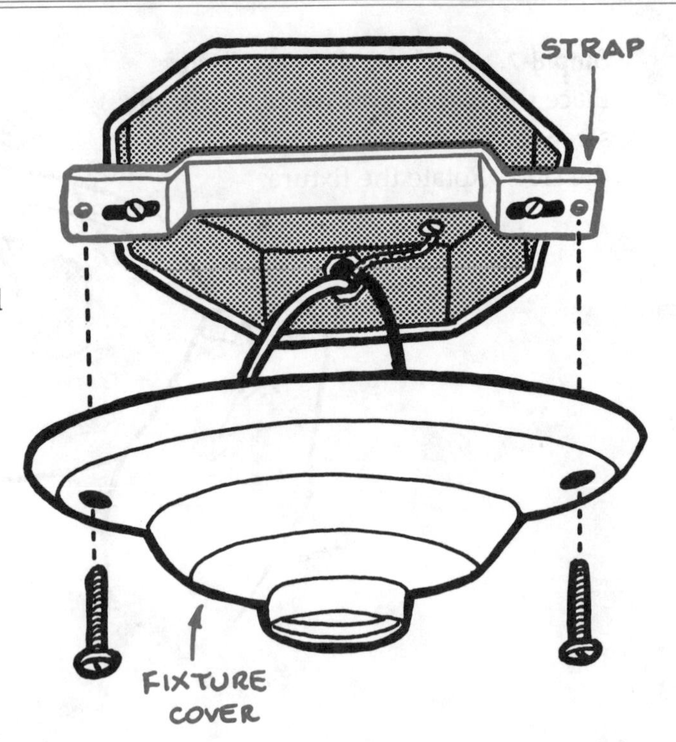


**Step 8-9. Strap-to-box mount.** Some ceiling fixtures are mounted to a strap attached to the box.

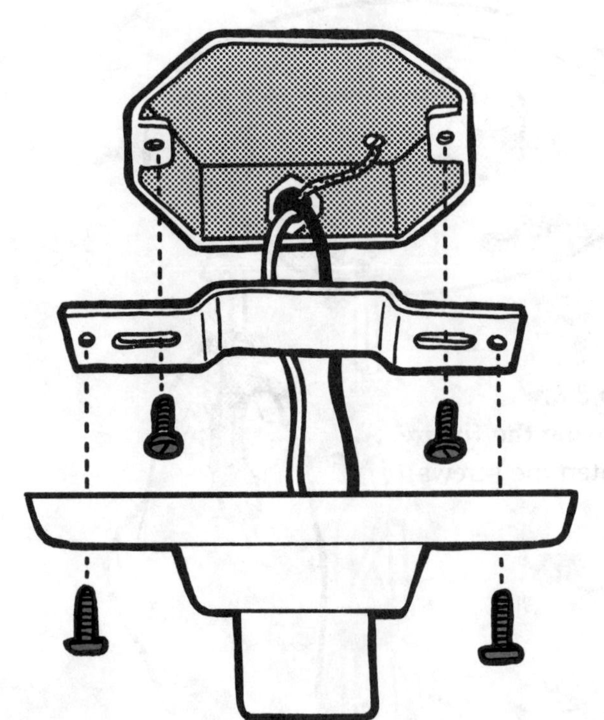

**Step 8-10. Replacing the fixture.** To remove this kind of fixture, remove the two screws from the fixture cover and the two from the strap. Lower the fixture and disconnect the wires. To install a new fixture, fasten the strap to the box so that it is centered over the box opening. Connect the wires and fasten the fixture to the strap.

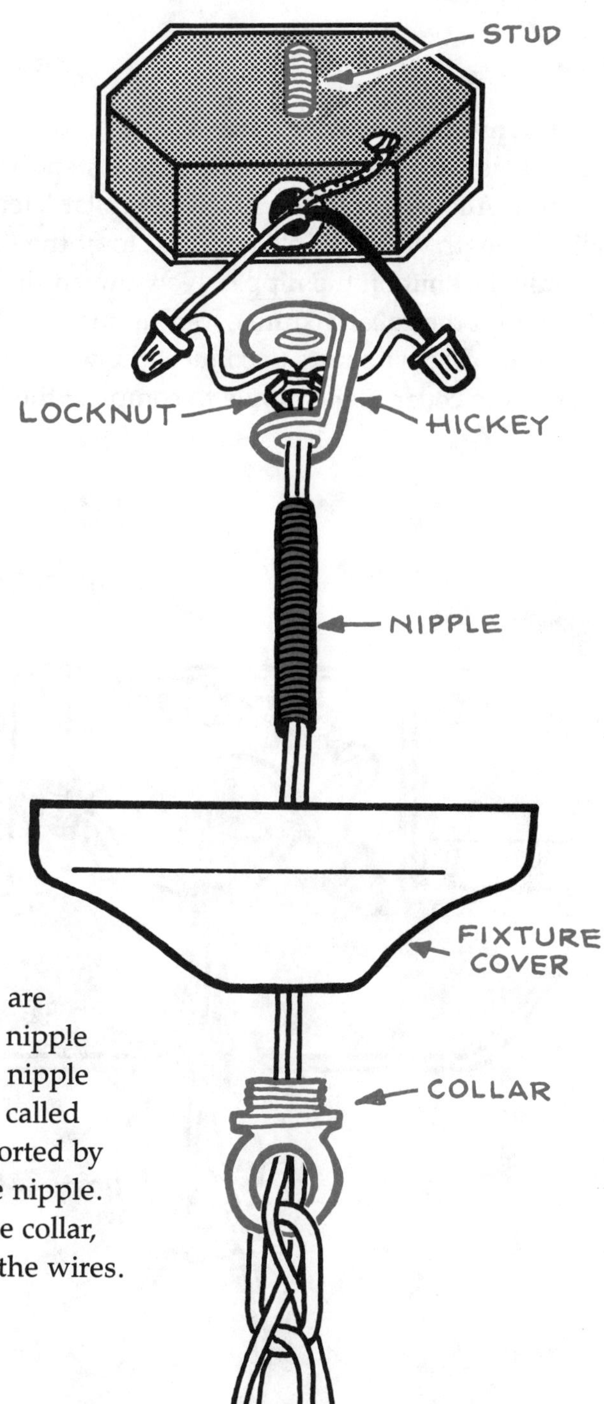


**Step 8-11.**
**Removing heavier fixtures.**
Heavier fixtures, like chandeliers, are sometimes held up by a threaded nipple attached to a stud in the box. The nipple is fastened to the stud by a fitting called a *hickey*. The fixture cover is supported by a collar threaded to the end of the nipple. To remove the fixture, unscrew the collar, lower the fixture, and disconnect the wires.

**Step 8-12. Installing heavier fixtures.**
To install a new fixture, use the hickey to fasten the nipple to the stud. Allowing room to attach the collar later, adjust the combined length of the nipple and stud to fit the fixture cover. Tighten the locknut on the nipple. Now thread the new fixture wires through the collar, fixture cover, nipple, and bottom half of the hickey. Use wire nuts to connect the wires. Install the cover and screw the collar to the nipple to complete the installation.

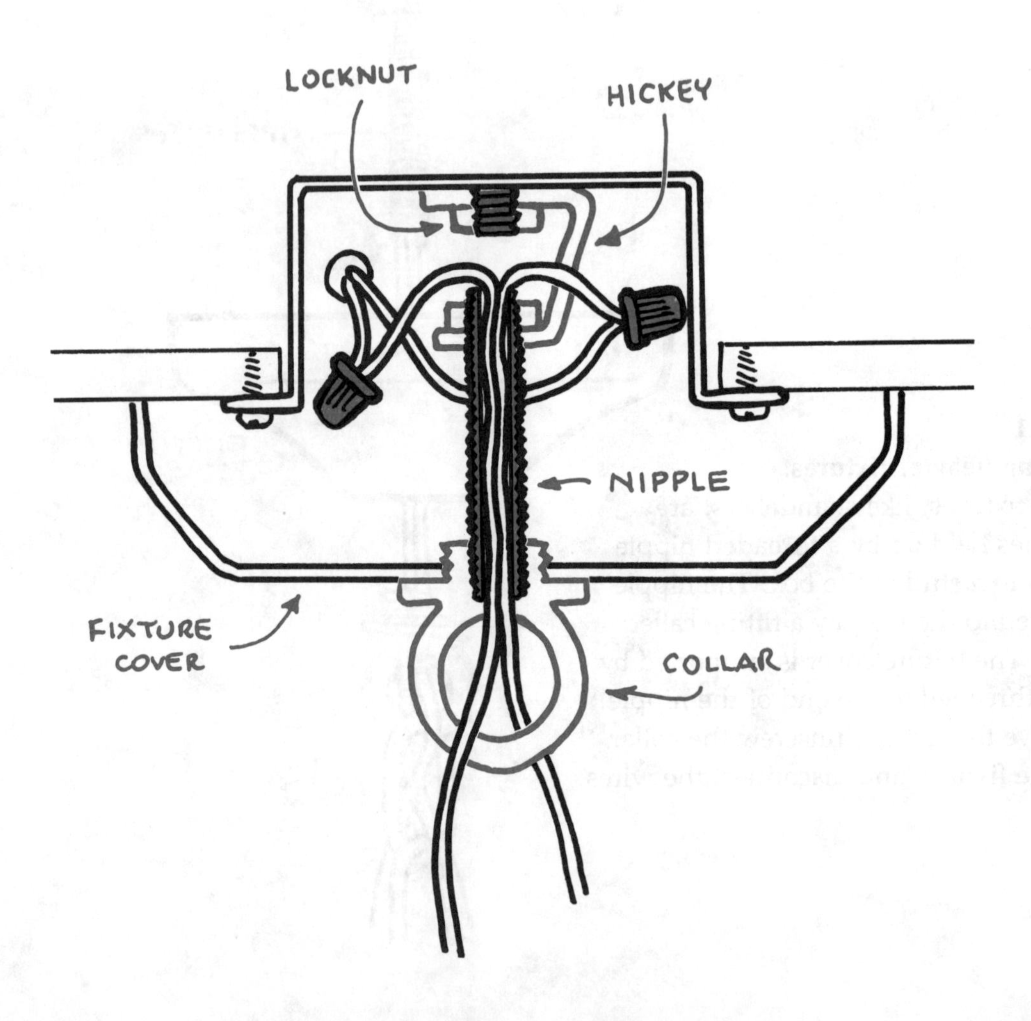

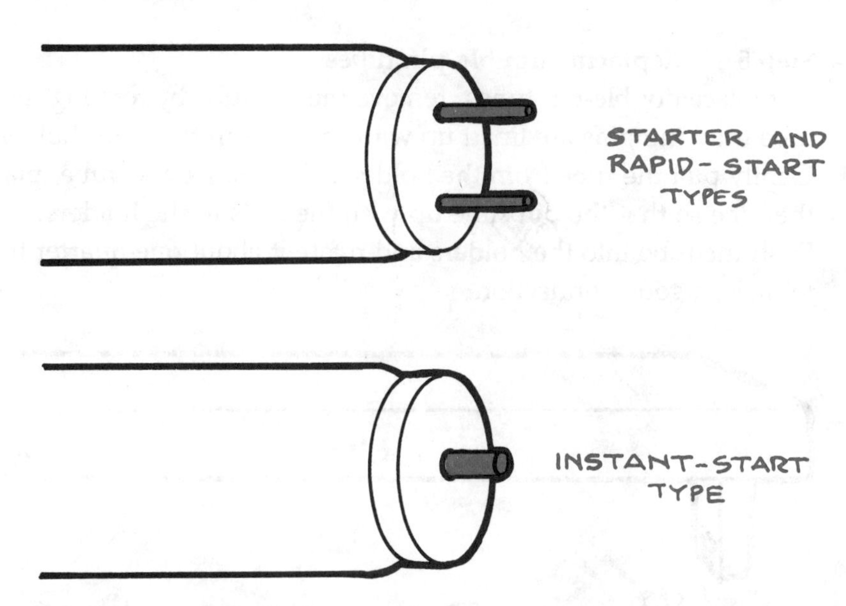


**Step 8-13. Fluorescent tubes.**
Three types of fluorescent lights are available: starter, rapid-start, and instant-start. The tubes for the starter and rapid-start type have two pins on each end. The instant-start type has only one pin.

**Step 8-14. Size and type.**
Markings near one end of the tube show the size and type of the tube.

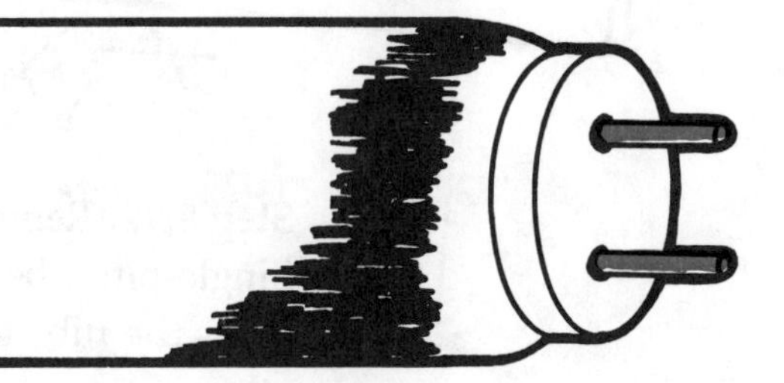

**Step 8-15. Blackened ends.**
Sometimes a gray band will appear about 2 inches from the end of the tube. This band indicates normal wear. If the ends are blackened, replace the tube.

**Step 8-16. Replacing double-pin tubes.**
To replace double-pin tubes, remove the old tube by rotating the tube until the pins are lined up with the slots in the lamp holders. Gently pull the tube from the holders. To install a new tube, place the tube so that the pins line up with the slots in the holders. Push the tube into the holders and rotate it about one-quarter turn to make a good connection.

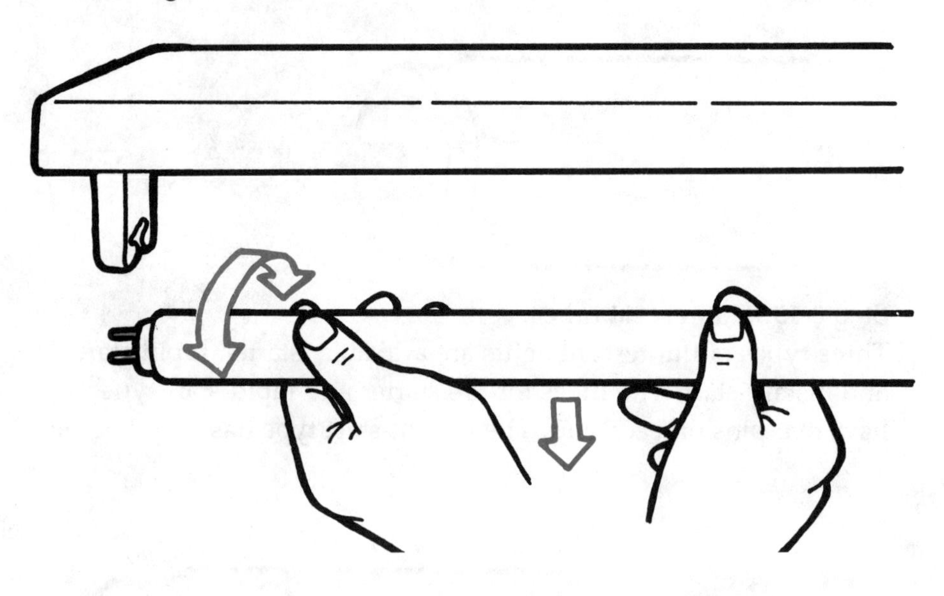

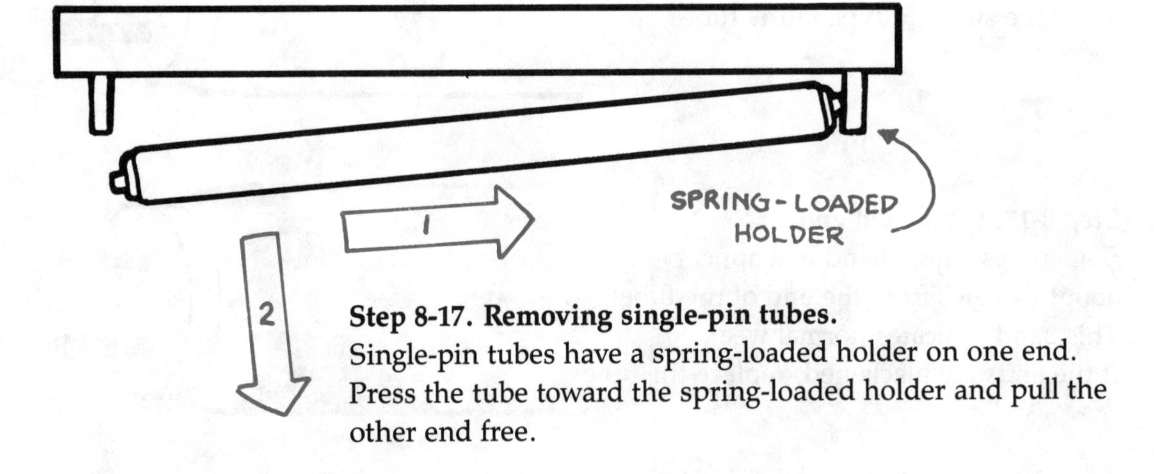


**Step 8-17. Removing single-pin tubes.**
Single-pin tubes have a spring-loaded holder on one end. Press the tube toward the spring-loaded holder and pull the other end free.

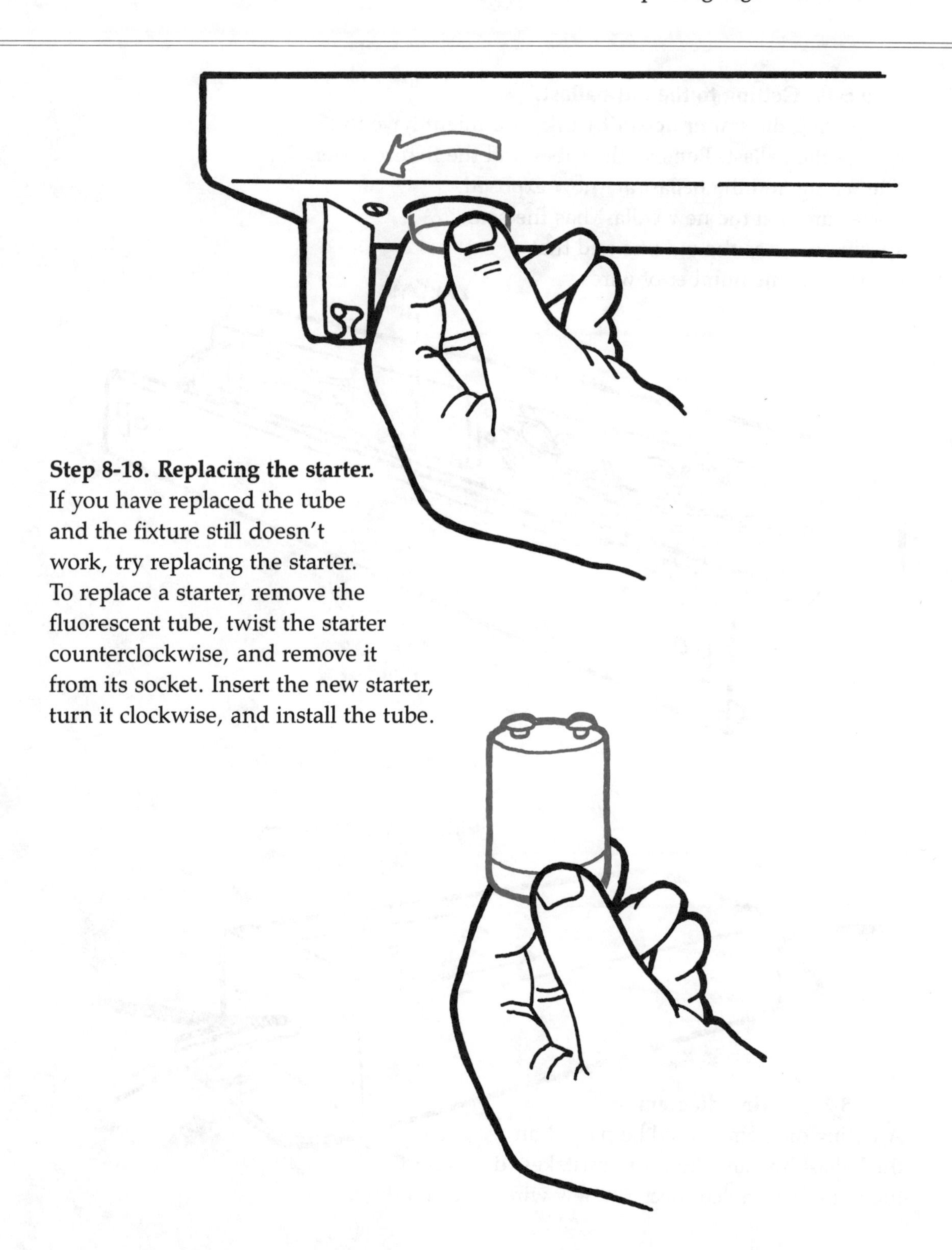

**Step 8-18. Replacing the starter.** If you have replaced the tube and the fixture still doesn't work, try replacing the starter. To replace a starter, remove the fluorescent tube, twist the starter counterclockwise, and remove it from its socket. Insert the new starter, turn it clockwise, and install the tube.

**Step 8-19. Getting to the old ballast.**
If replacing the starter doesn't work, you might have to replace the ballast. Remove the tubes and the fixture cover. The wiring and the ballast are now exposed. Make sure that the new ballast has the same specifications as the old one and that it has the same number of wires.

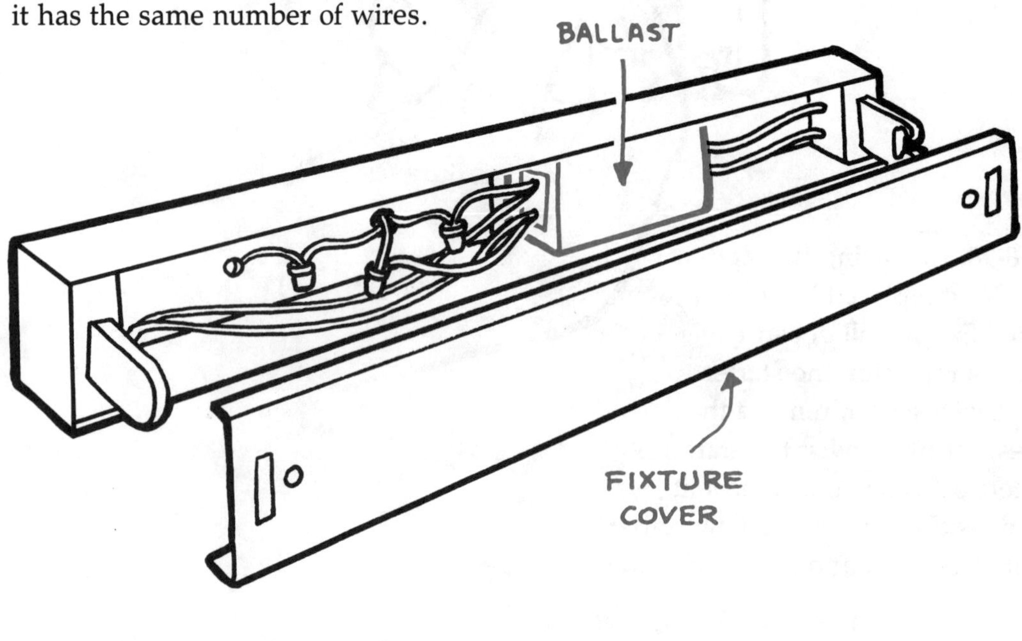


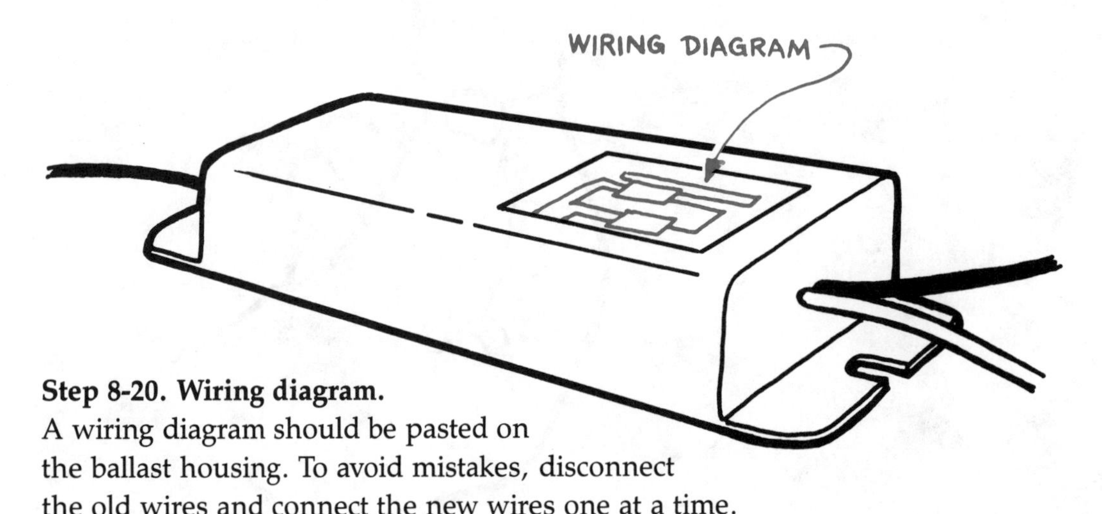


**Step 8-20. Wiring diagram.**
A wiring diagram should be pasted on the ballast housing. To avoid mistakes, disconnect the old wires and connect the new wires one at a time.

**Step 8-21. Installing the new ballast.**
After the connections are made, loosen the screws holding the old ballast and install the new one. Replace the fixture cover and reinstall the tubes.

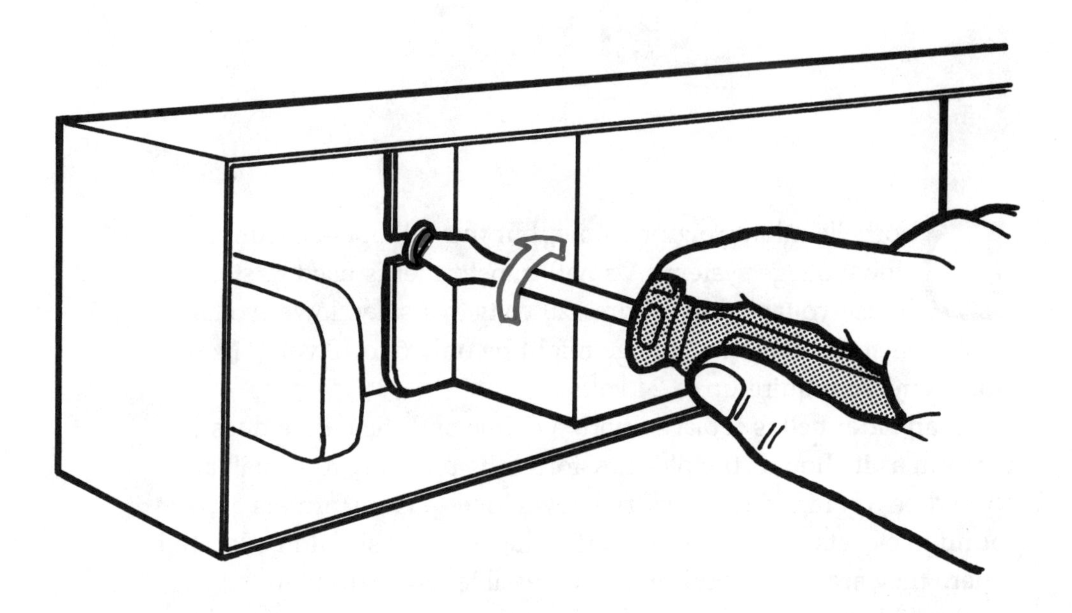

CHAPTER NINE

# Replacing Doorbells & Chimes

Doorbells might ring or chime, but they all operate from a low-voltage system. A small transformer is used to step the house voltage down from 120 volts to a safer, lower voltage. In older installations, the voltage might be only 6 to 10 volts. Newer chimes might require up to 24 volts.

If an older bell is replaced with a chime and the chime doesn't perform as it should, the old transformer is probably too small and should be upgraded to match the new chimes. Transformers are found in closets, utility rooms, and garages. They should be located where they are visible and readily accessible, away from any combustible material. Do not, for example, place a transformer in an attic, where any problem might go unnoticed for some time.

## Tools & Materials

- ☐ Screwdriver
- ☐ Wire stripper
- ☐ Volt-ohmmeter
- ☐ New chimes
- ☐ New transformer (if necessary)

Before replacing faulty chimes, check them out. Often chimes will not sound because they are gummed up with dirt and lint. When troubleshooting doorbell circuits, you normally have to leave the power on. Turn off the power if you are doing any work on the 120-volt side of the transformer.

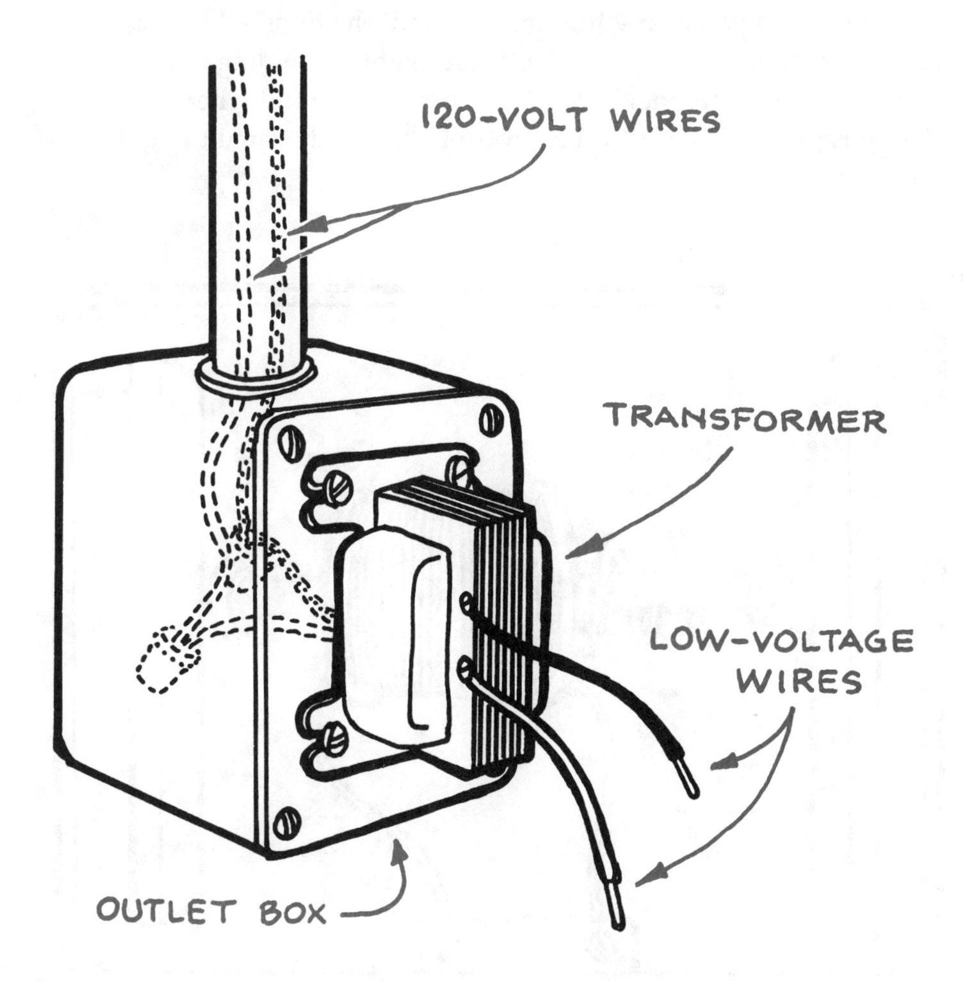


**Step 9-1. Checking for low voltage.**
First use the volt-ohmmeter to check if low voltage is coming out of the transformer. If not, the breaker might be tripped or the transformer might be bad.

**Step 9-2. Checking chimes.**
If you verify low voltage, go to the chimes. Have someone press the doorbell. If you hear a humming sound, the doorbell button itself is probably okay. The striker shaft might be binding and, therefore, need cleaning. If you don't hear any sound when the button is pressed, the button is probably bad or the circuit might have a break.

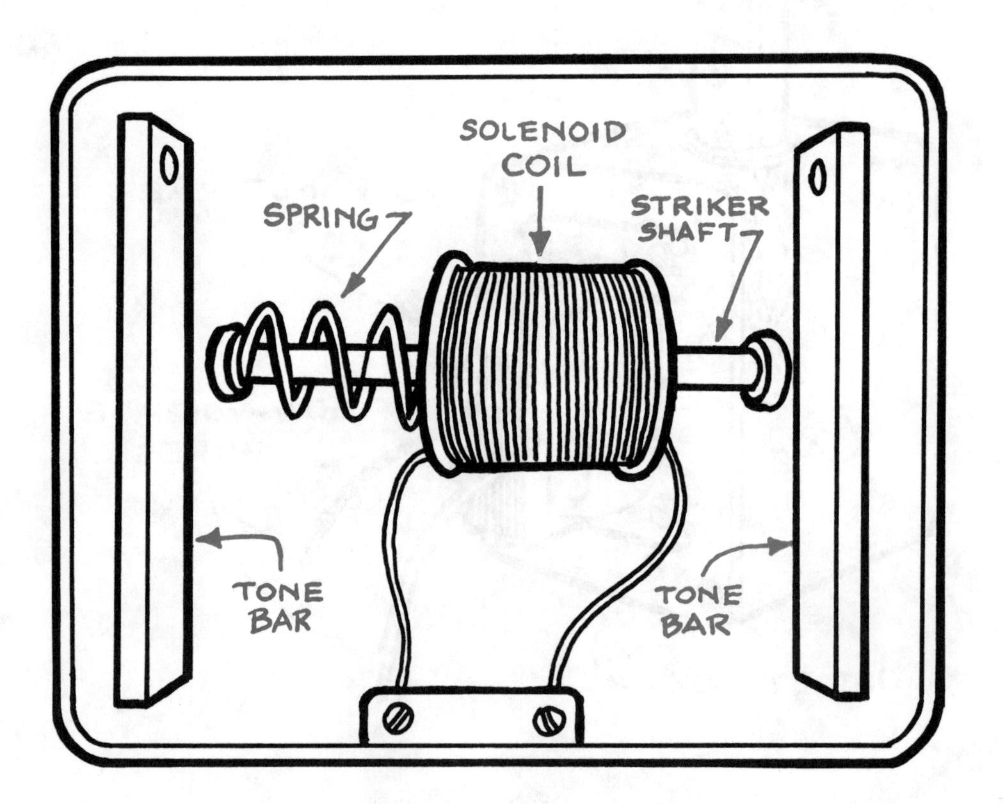

**Step 9-3. Testing the push button.**
To check out the button, remove the cover and the mounting screws. Disconnect the two wires and touch them together. If the chime sounds, replace or repair the button. Often, just cleaning the striker mechanism or tightening an electrical connection will solve the problem.

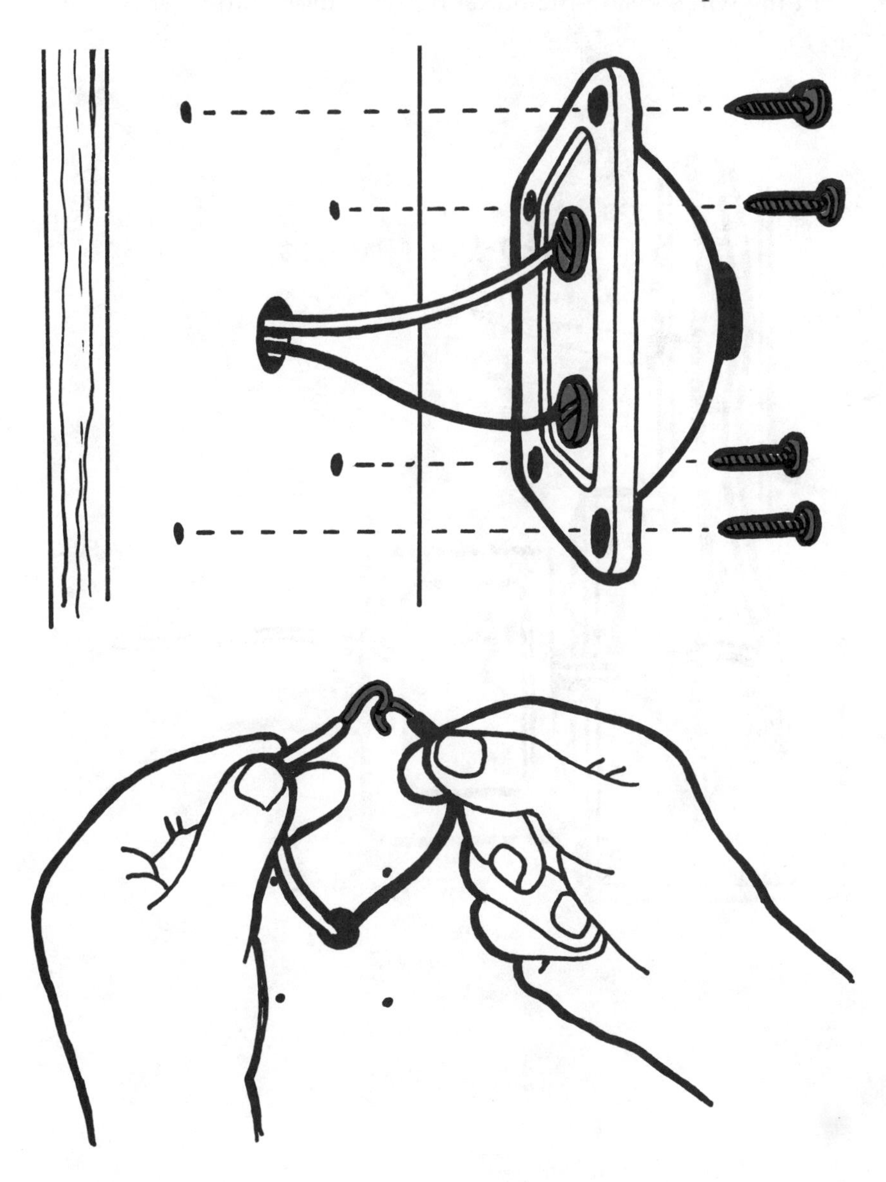

**Step 9-4. Installing a new transformer.**
Turn off the power and remove the old transformer, noting how the wire connections were made. Disconnect the house wires from the old transformer and reconnect them to the new one. Fold the wires inside the outlet box and mount the transformer to the box.

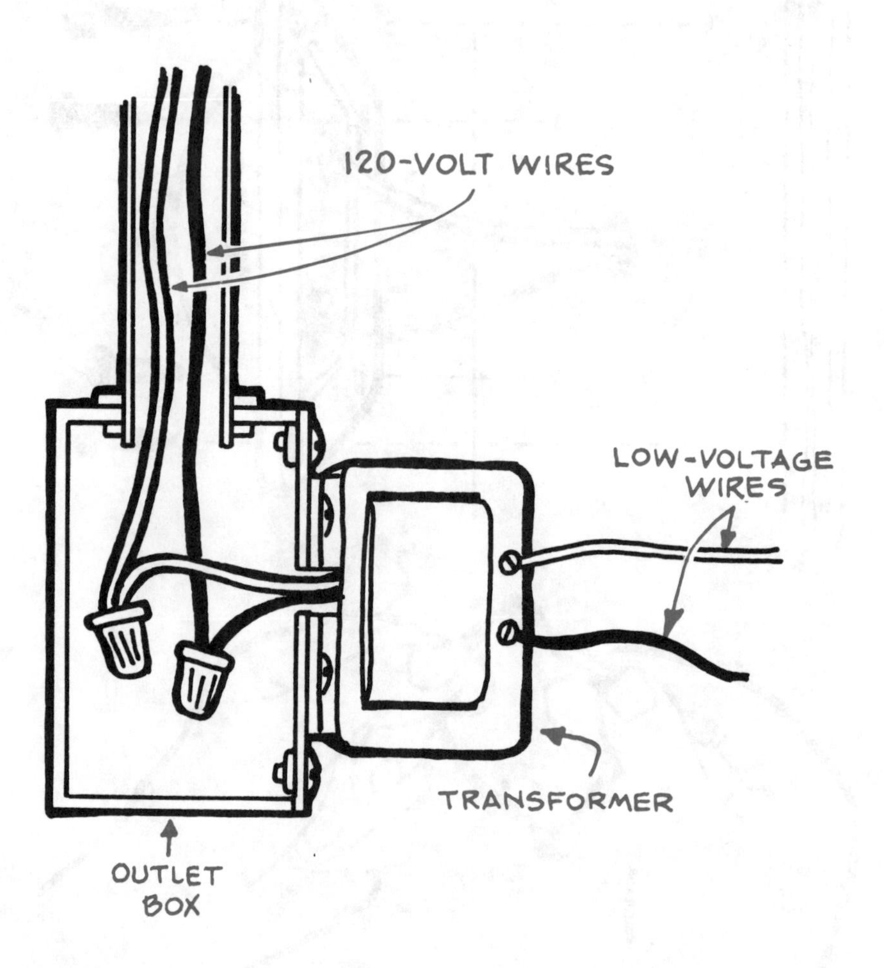

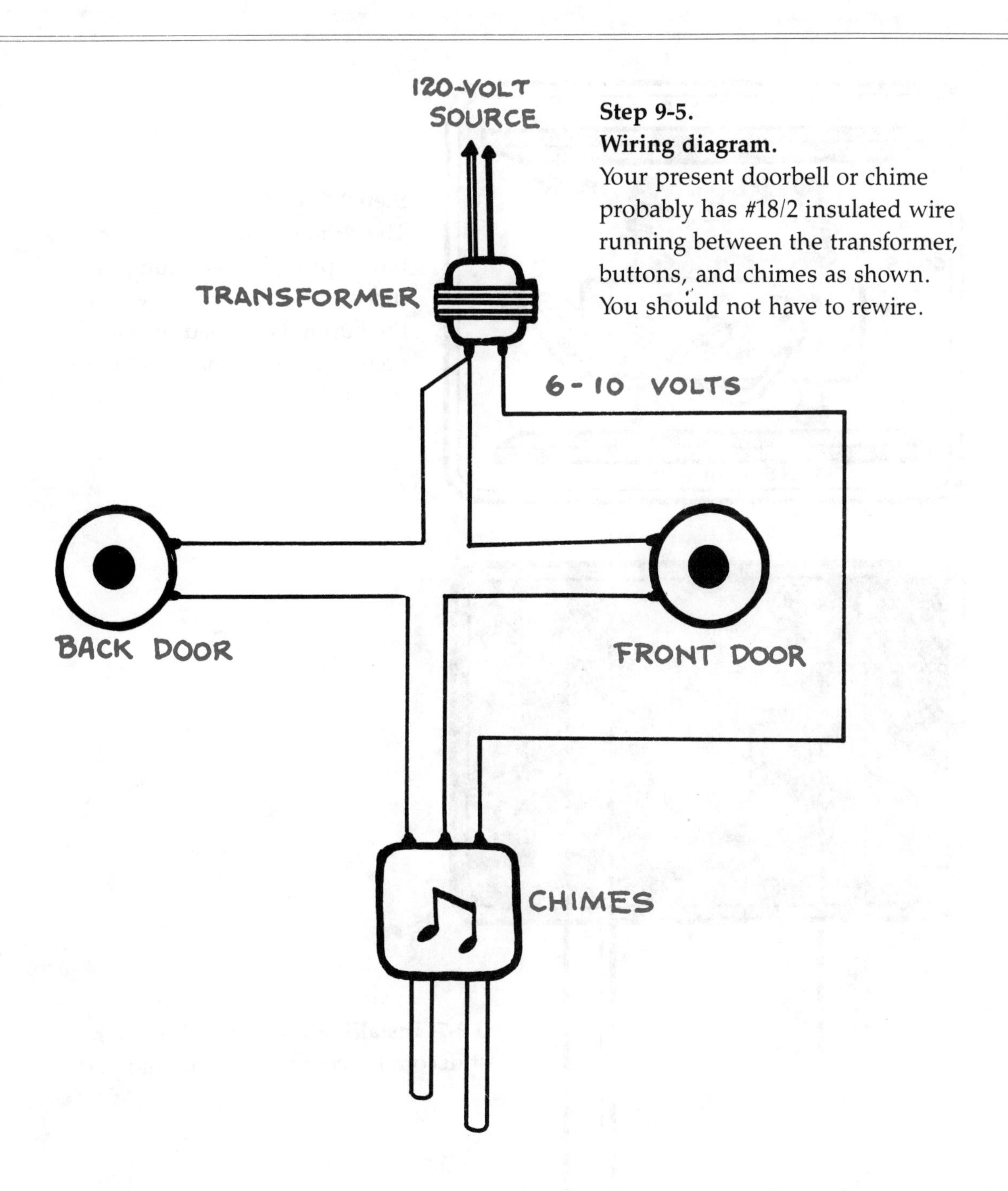


**Step 9-5.**
**Wiring diagram.**
Your present doorbell or chime probably has #18/2 insulated wire running between the transformer, buttons, and chimes as shown. You should not have to rewire.

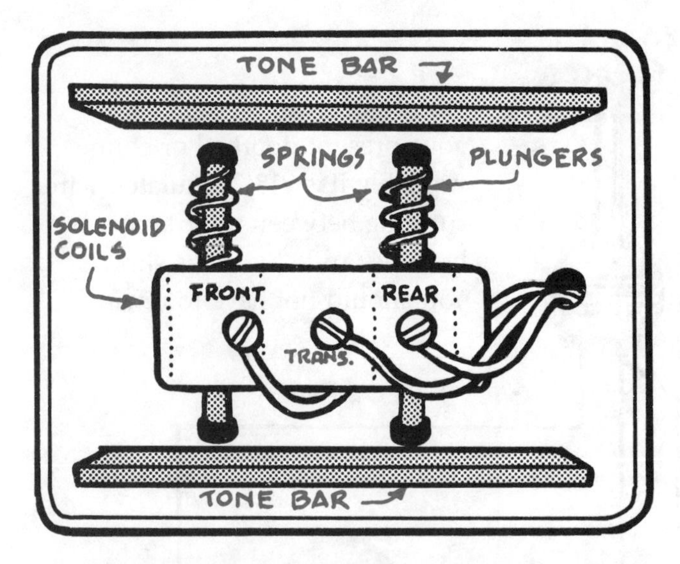


**Step 9-6. Chimes.**
The chimes consist of two or more spring-loaded plungers that strike a bar or tube when the button is pushed. Install according to the manufacturer's instructions.

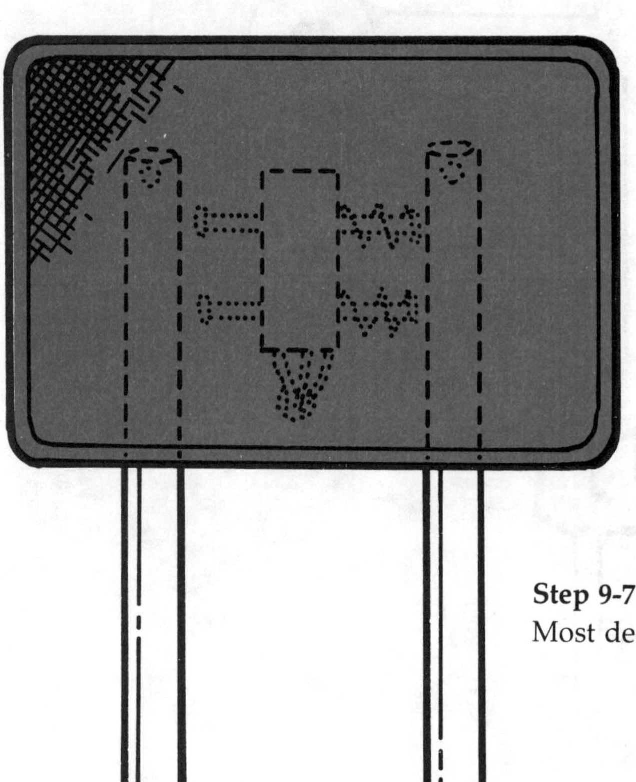

**Step 9-7. Installing the decorative cover.**
Most decorative covers just snap into place.

**Step 9-8.**
**Connecting the push button.**
The button has two terminal screws.
The wires can be connected to either screw.

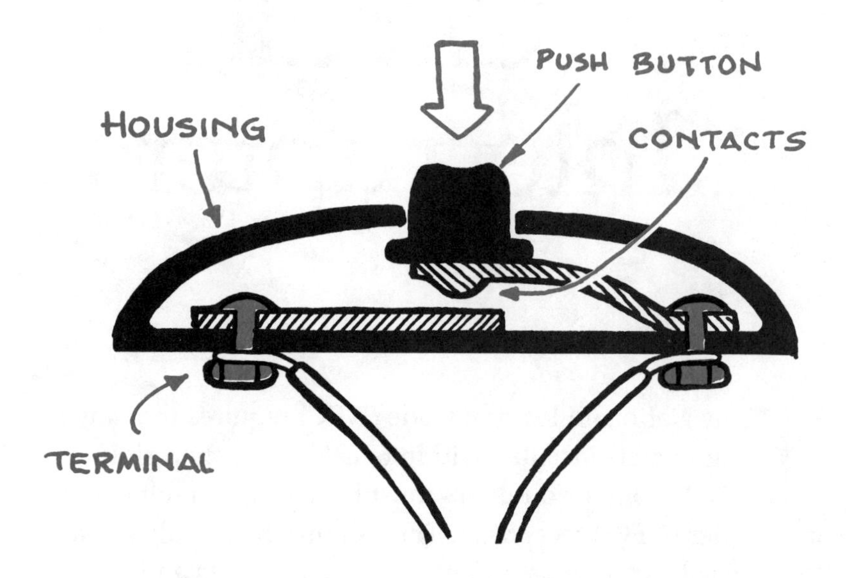

CHAPTER TEN

# Installing Ground-fault Circuit Interrupters

The National Electrical Code (NEC) requires that any new receptacle installed within 6 feet of a kitchen sink, in a bathroom, or outdoors must be a ground-fault circuit interrupter (GFCI) receptacle. They should be installed in any location subject to moisture, including basements and garages. This regulation applies specifically to receptacles that you use intermittently. Receptacles serving appliances that are never unplugged, such as freezers or garage door openers, do not have to be GFCIs.

A GFCI can be installed in a regular receptacle box. It has "test" and "reset" buttons that should be used about once a month to ensure that the device is working. Most GFCIs have provisions to connect to other regular receptacles, providing GFCI protection to any receptacle installed downstream in the circuit.

## Tools & Materials

- ☐ Screwdriver
- ☐ Needle-nose pliers
- ☐ Wire stripper
- ☐ Wire nuts
- ☐ Neon voltage tester or volt-ohmmeter
- ☐ New GFCI receptacle

To install a GFCI receptacle, first make sure the power to the circuit is off. Remove the receptacle cover plate and unscrew the two mounting screws.

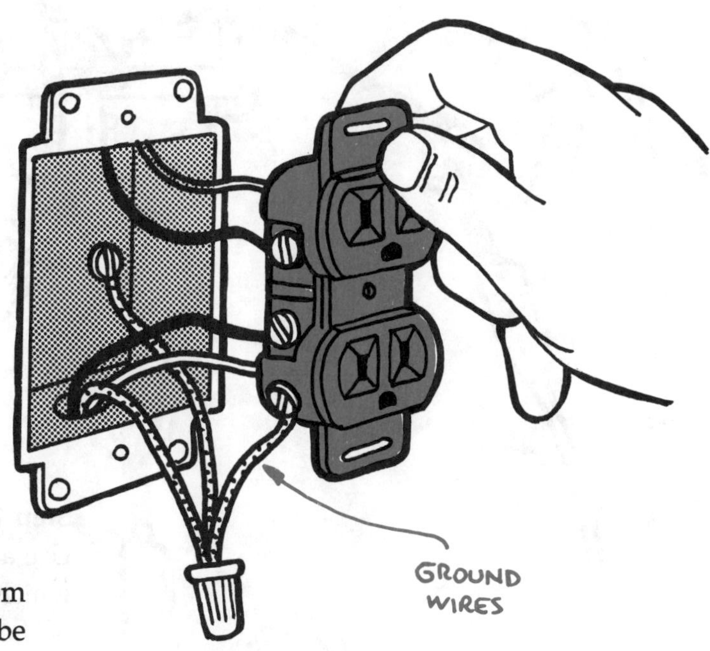


**Step 10-1.**
**Removing the old receptacle.**
Carefully pull the receptacle from the box. Recheck for voltage to be sure that the power is off. You should see black, white, and ground wires just like any receptacle. The receptacle might have two sets of wires: one set from the service panel, the other cable feeding the receptacles downstream from this one. Disconnect the wires from the receptacle and bend them out of the way. If two sets of wires are present, you will need to determine which set comes from the service panel (the hot wire). Make sure the wires are clear, and turn the power back on.

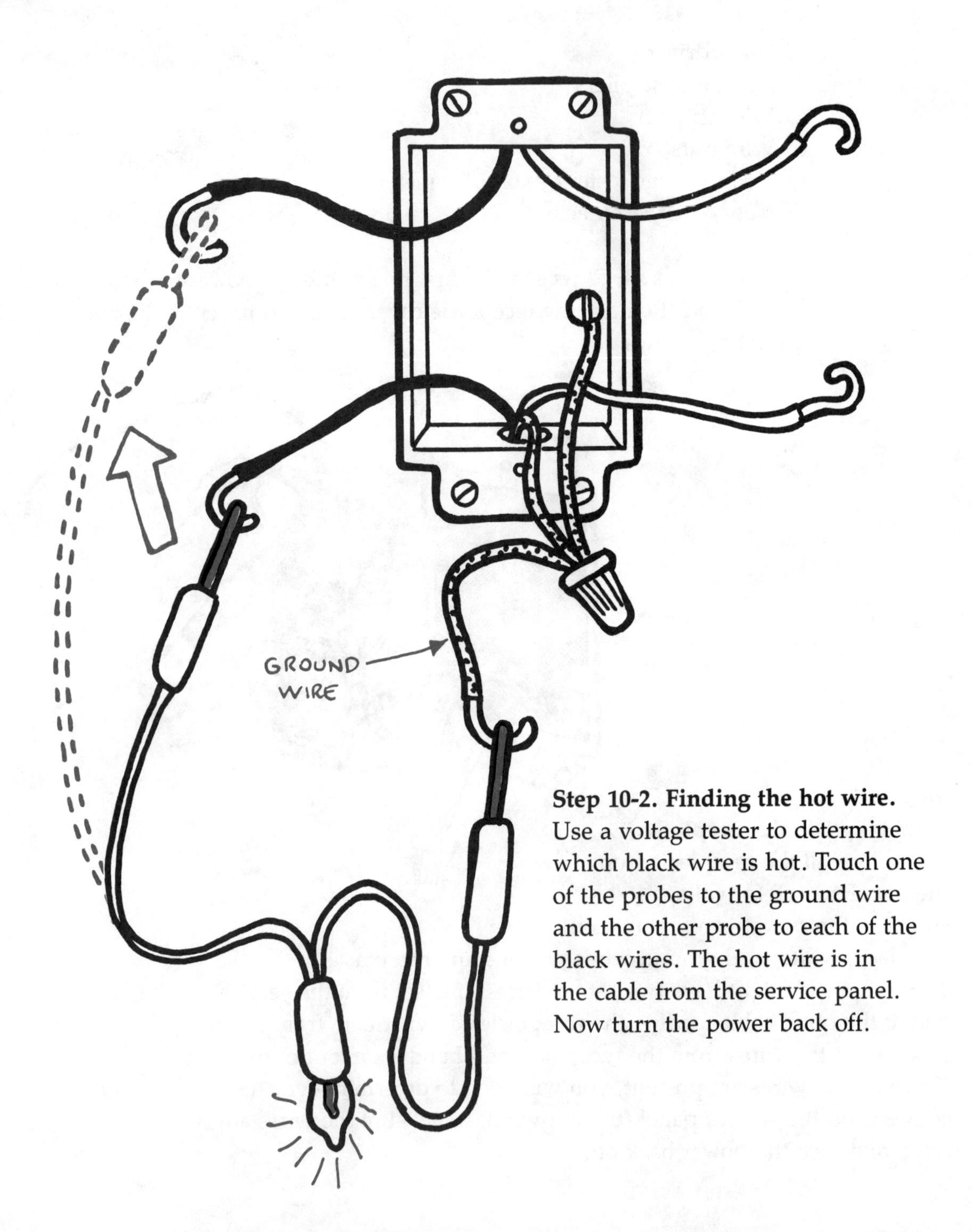


**Step 10-2. Finding the hot wire.** Use a voltage tester to determine which black wire is hot. Touch one of the probes to the ground wire and the other probe to each of the black wires. The hot wire is in the cable from the service panel. Now turn the power back off.

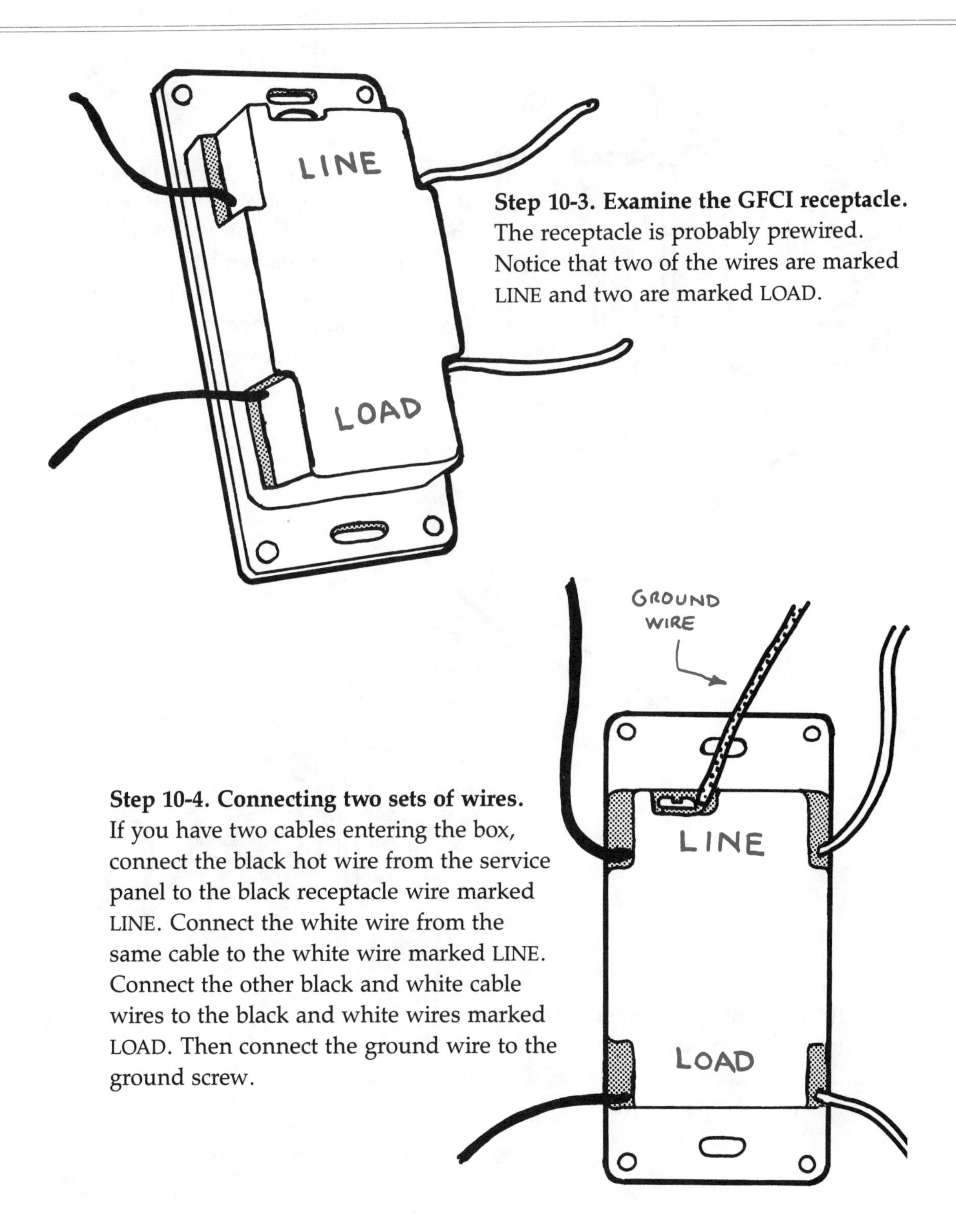


**Step 10-3. Examine the GFCI receptacle.** The receptacle is probably prewired. Notice that two of the wires are marked LINE and two are marked LOAD.

**Step 10-4. Connecting two sets of wires.** If you have two cables entering the box, connect the black hot wire from the service panel to the black receptacle wire marked LINE. Connect the white wire from the same cable to the white wire marked LINE. Connect the other black and white cable wires to the black and white wires marked LOAD. Then connect the ground wire to the ground screw.

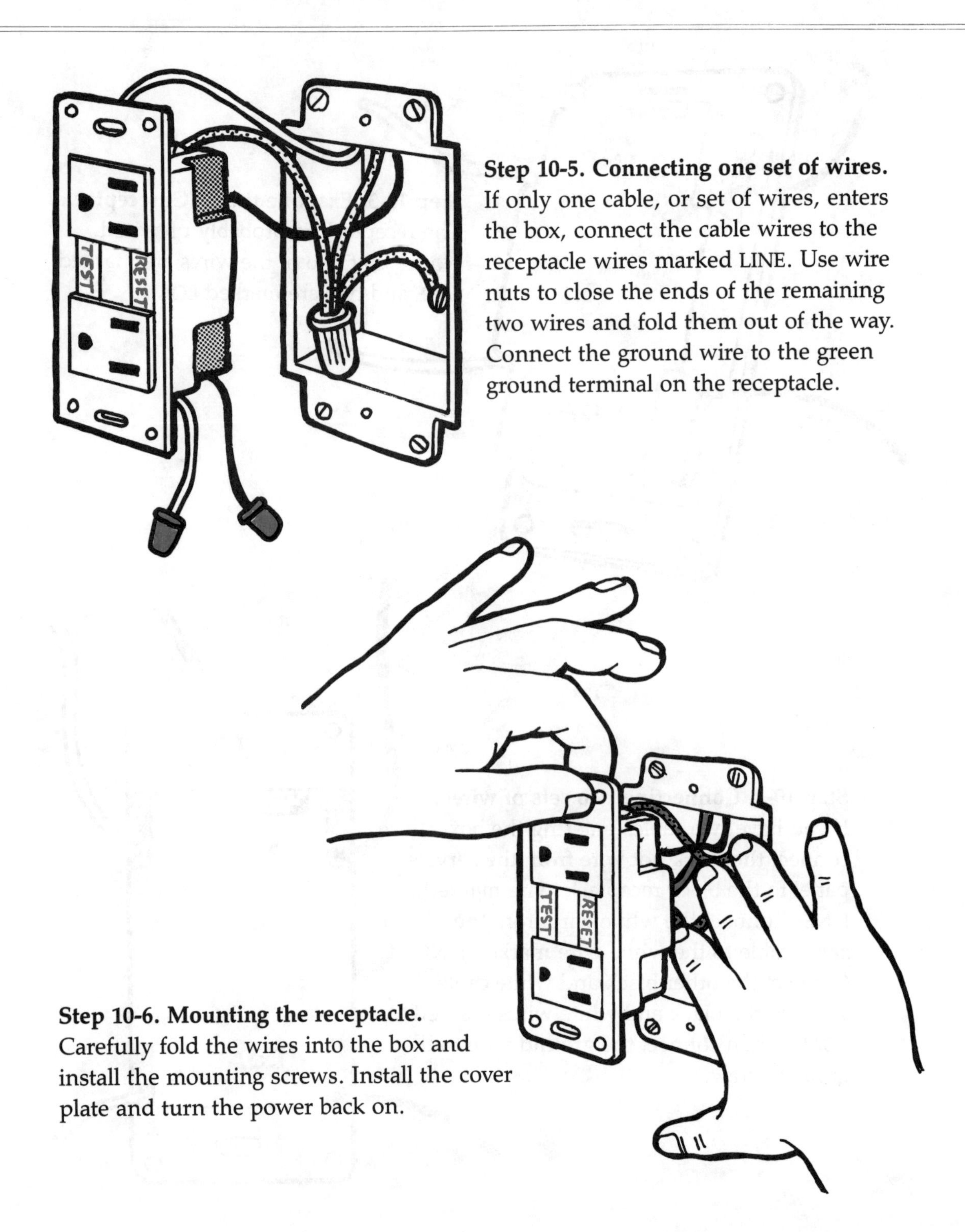


**Step 10-5. Connecting one set of wires.** If only one cable, or set of wires, enters the box, connect the cable wires to the receptacle wires marked LINE. Use wire nuts to close the ends of the remaining two wires and fold them out of the way. Connect the ground wire to the green ground terminal on the receptacle.

**Step 10-6. Mounting the receptacle.** Carefully fold the wires into the box and install the mounting screws. Install the cover plate and turn the power back on.

CHAPTER ELEVEN

# Installing Dimmer Switches

Living and dining areas often have overhead lights that are too bright for all situations. Dimmer switches allow you to vary the intensity to suit the mood of the occasion. Most dimmer switches, however, can be used only with incandescent lights, the regular light bulbs commonly used in the home. Special dimmer switches are required for fluorescent lights. Dimmer switches have a control knob that allows you to adjust the light from bright to dim or anywhere in between. They have the additional advantage of reducing electrical consumption when turned to any point below full bright.

## Tools & Materials

- ☐ Screwdriver
- ☐ Needle-nose pliers
- ☐ Wire stripper
- ☐ Neon voltage tester or volt-ohmmeter
- ☐ Dimmer switch

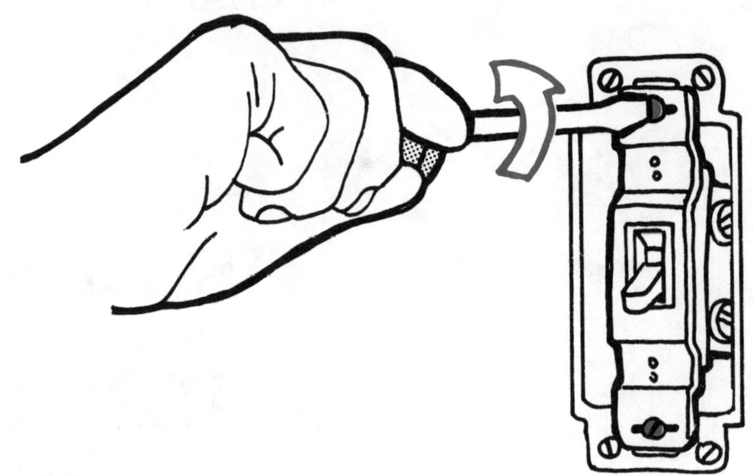

**Step 11-1. Remove the old switch.**
To install a dimmer switch, first turn off the power to the circuit. Remove and disconnect the old switch.

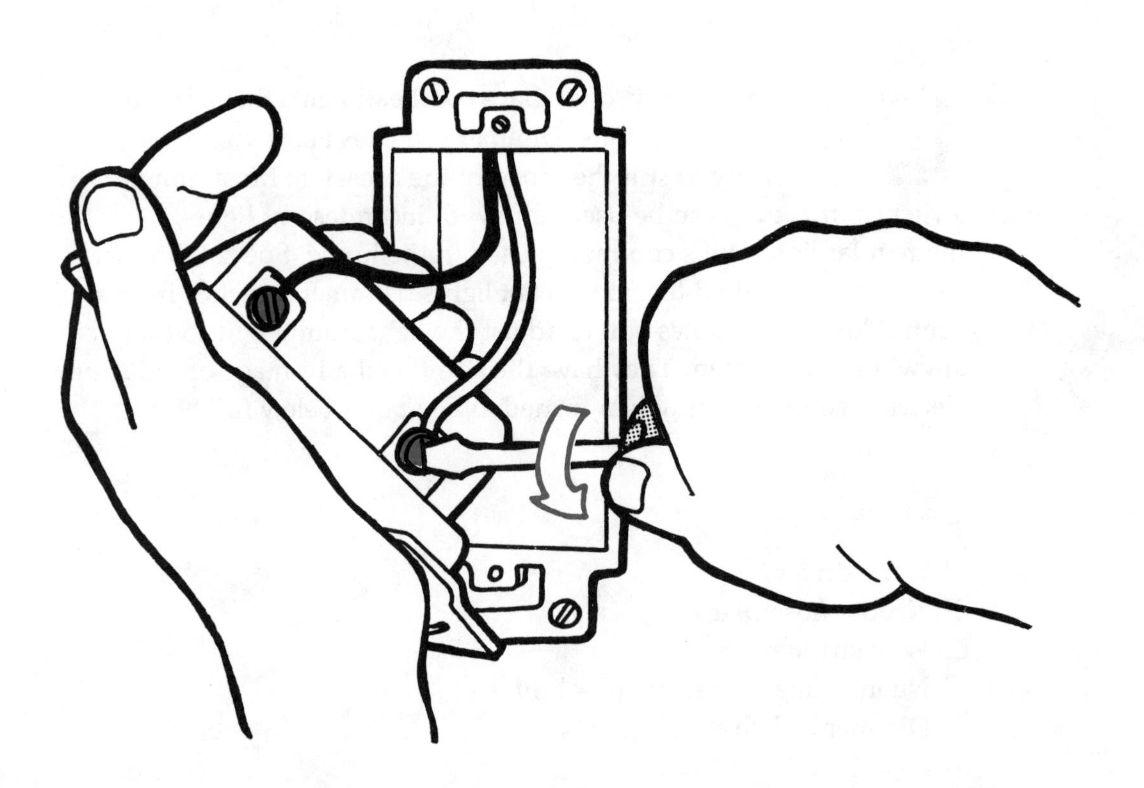

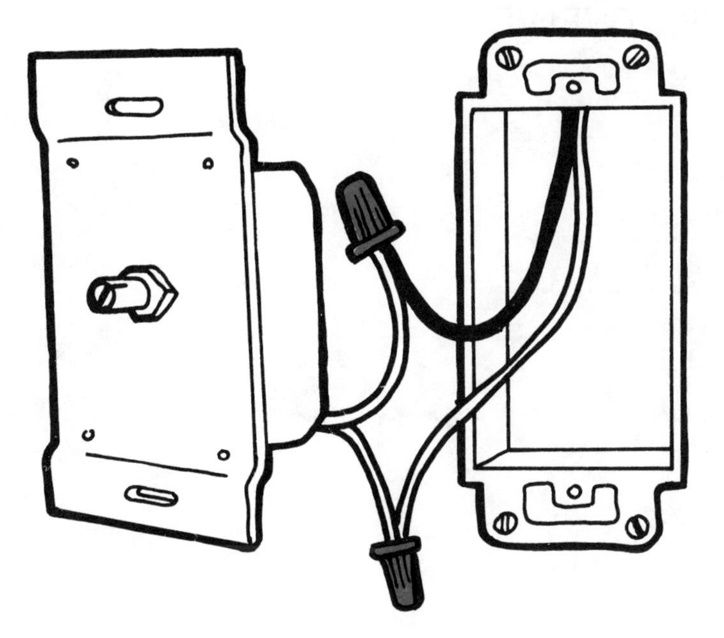

**Step 11-2.**
**Connecting the wires.**
Notice that the dimmer switch has two wires coming from it. Use wire nuts to connect the house wires to the two wires on the new switch. Either switch wire can be connected to either house wire.

**Step 11-3. Mounting the switch.**
Carefully fold the wires into the box and mount the switch. Now mount the cover plate, install the control knob, and turn the power back on.

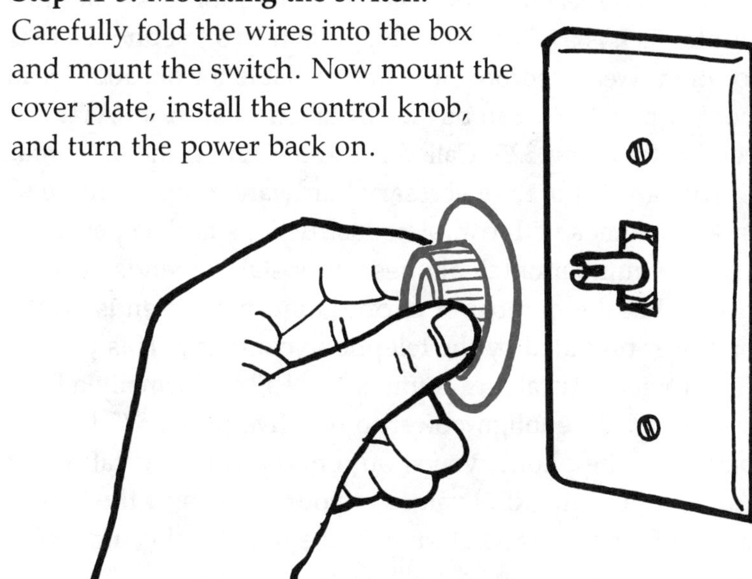

CHAPTER TWELVE

# Adding Telephone Jacks

Telephone jacks usually are installed in homes during construction, but these outlets are often too few or in inconvenient locations. Wall jacks come in a variety of sizes and configurations. They are sometimes called PMCs, or plug-in modular connectors. They can be surface- or flush-mounted. A duplex jack is available for plugging in two telephones at the same location. Weatherproof jacks are available for outdoor locations by the pool or patio. You can buy an installation kit from AT&T telephone centers for about $25. Called Add-An-Outlet, the kit contains 50 feet of wire and all of the necessary hardware to install three telephone jacks. It comes with complete instructions and a special wire stripper.

The number of telephones you install depends on your personal needs, but the number of phones that ring depends on the amount of power provided by the telephone company. This power is called the "ringer equivalence number" or REN. Normally a home is supplied with enough power to ring five phones (5 REN). On the bottom of the phone, you should find a label indicating the REN. Add up all of the REN values of your phones. If the total doesn't exceed 5, they should all ring. If the total REN is more than 5, all the phones will work but not all of them will ring.

In some areas, the service supplied to the home is designed for rotary phones. Most phones currently on the market have selector switches to switch from touch-tone to rotary systems. If your home is supplied by a rotary system and your touch-tone phone does not have the selector switch, contact your local phone company and ask them to upgrade your service.

Voltages powering a home telephone service can vary. While normally these voltages are not dangerous, you should take some precautions. Anyone with a pacemaker should not work on a telephone system connected to the incoming service. If the telephone happens to ring, the voltage can increase dramatically. Before making connections, lift one of the existing phones off the receiver. Then make the connections and replace the phone.

## Tools & Materials

- ☐ Screwdriver
- ☐ Wire stripper
- ☐ Staple gun
- ☐ Hammer
- ☐ Telephone cable
- ☐ Telephone jack

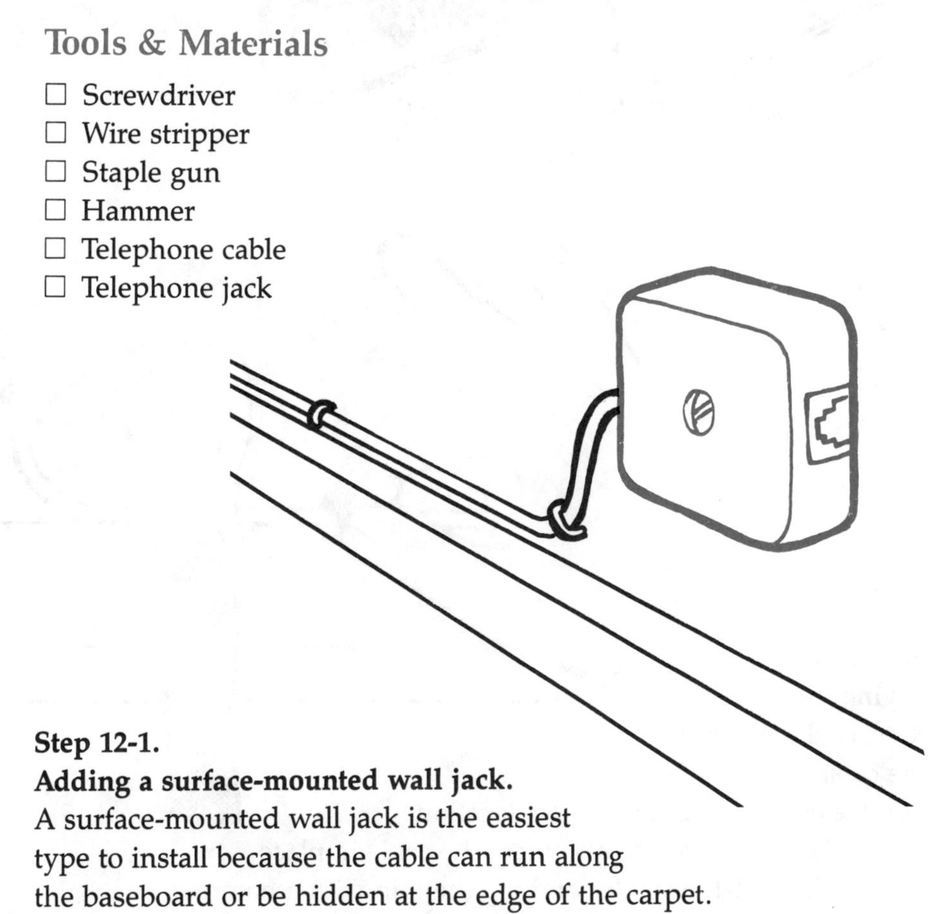

**Step 12-1.**
**Adding a surface-mounted wall jack.**
A surface-mounted wall jack is the easiest type to install because the cable can run along the baseboard or be hidden at the edge of the carpet.

**Step 12-2. Connecting new wires.**
A standard cable has four wires: yellow, black, green, and red. Remove the cover on the existing wall jack. Examine the terminal screws. You will see letters by each screw indicating which colored wire connects to which screw. Loosen the screws and connect the new wires to the terminals according to the color code.

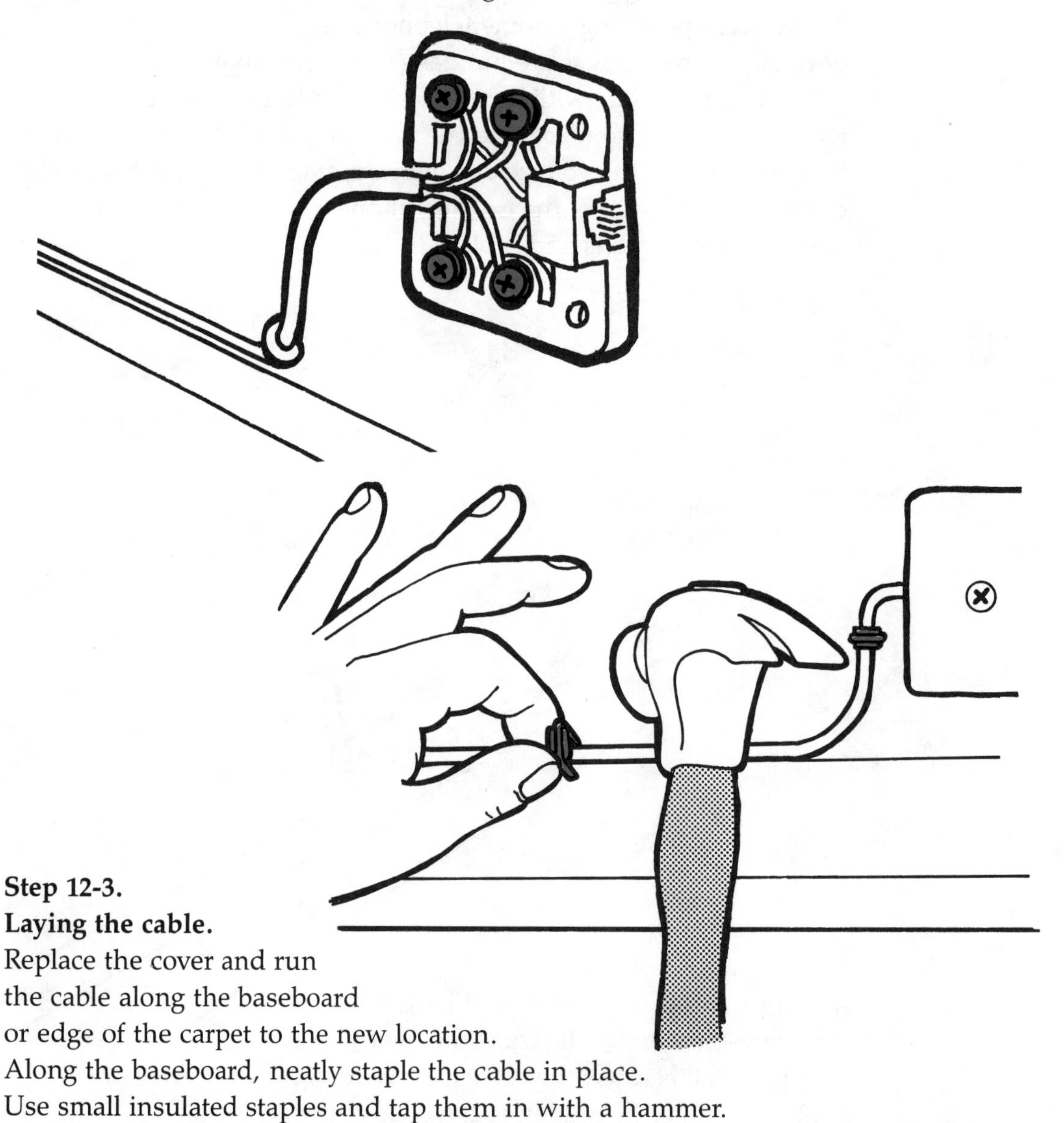

**Step 12-3. Laying the cable.**
Replace the cover and run the cable along the baseboard or edge of the carpet to the new location. Along the baseboard, neatly staple the cable in place. Use small insulated staples and tap them in with a hammer.

**Step 12-4. Fishing for the cable.**
If you need to go through a wall, place the blade of a small screwdriver against the wall and tap the end gently with a hammer to make a small hole. If you don't run into an obstruction such as a stud, do the same on the other side of the wall. Use a small wire to fish the cable through the hole.

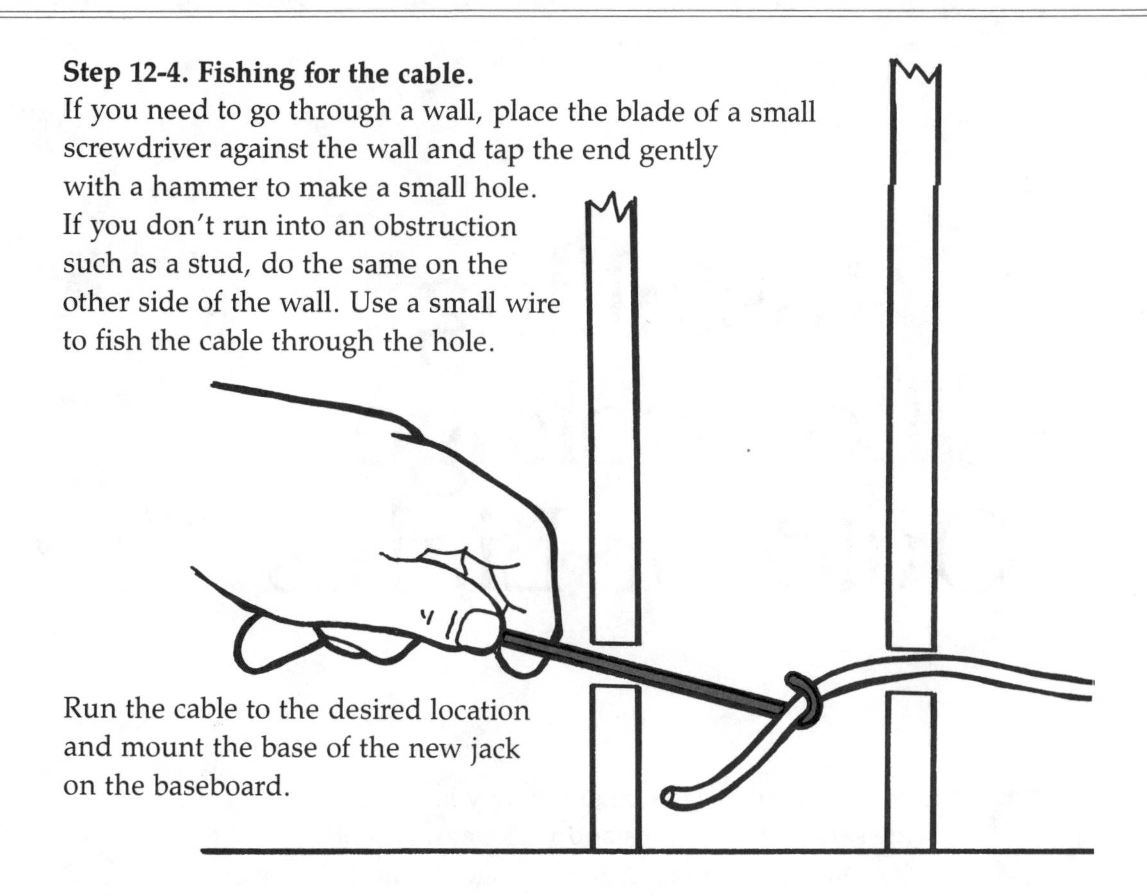

Run the cable to the desired location and mount the base of the new jack on the baseboard.

**Step 12-5. Connecting the new jack.**
Connect the wires to their proper terminal screws and install the cover of the jack. Plug in the phone and test the installation.

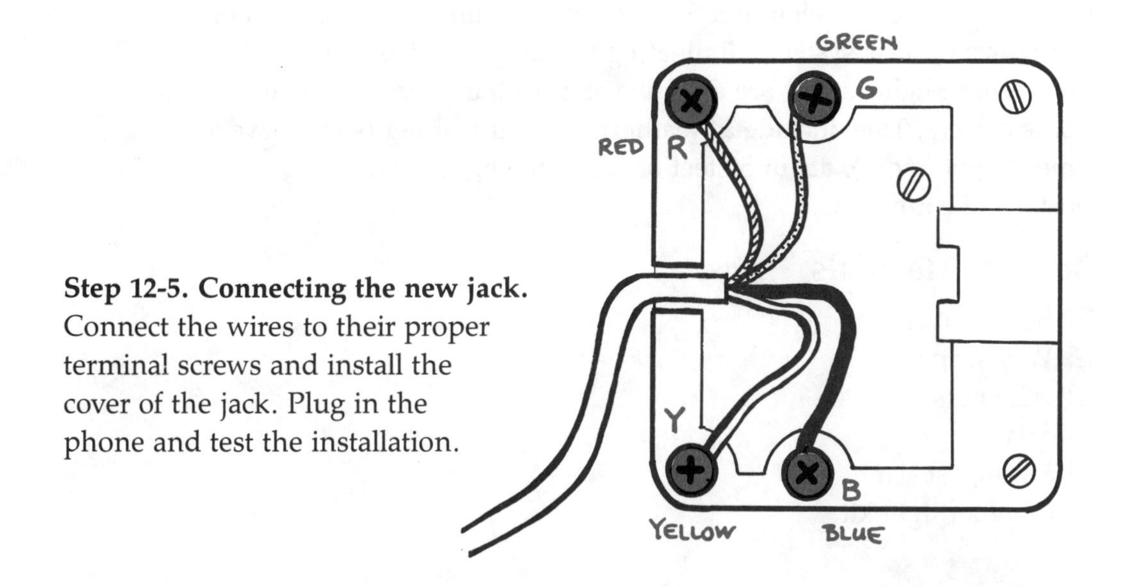

CHAPTER THIRTEEN

# Installing Low-voltage Outdoor Lights

Outdoor lights enhance the beauty of any landscape, provide safe travel on steps and walkways, and offer security against intruders. A low-voltage circuit makes a very simple and affordable installation. A typical system might have 4, 6, or 8 lights and cost well under $100. It operates on 12 volts instead of the 120-volt house system, eliminating the danger of harmful shocks. Low-voltage lighting kits are available at most hardware stores and home centers. They include a weatherproof transformer (some have a timer or photo cell), about 50 feet of cable, the lights, mounting posts, and connectors.

## Tools & Materials

- ☐ Screwdriver
- ☐ Wire stripper
- ☐ Wire nuts
- ☐ Spade
- ☐ Silicone sealant
- ☐ Outdoor light kit

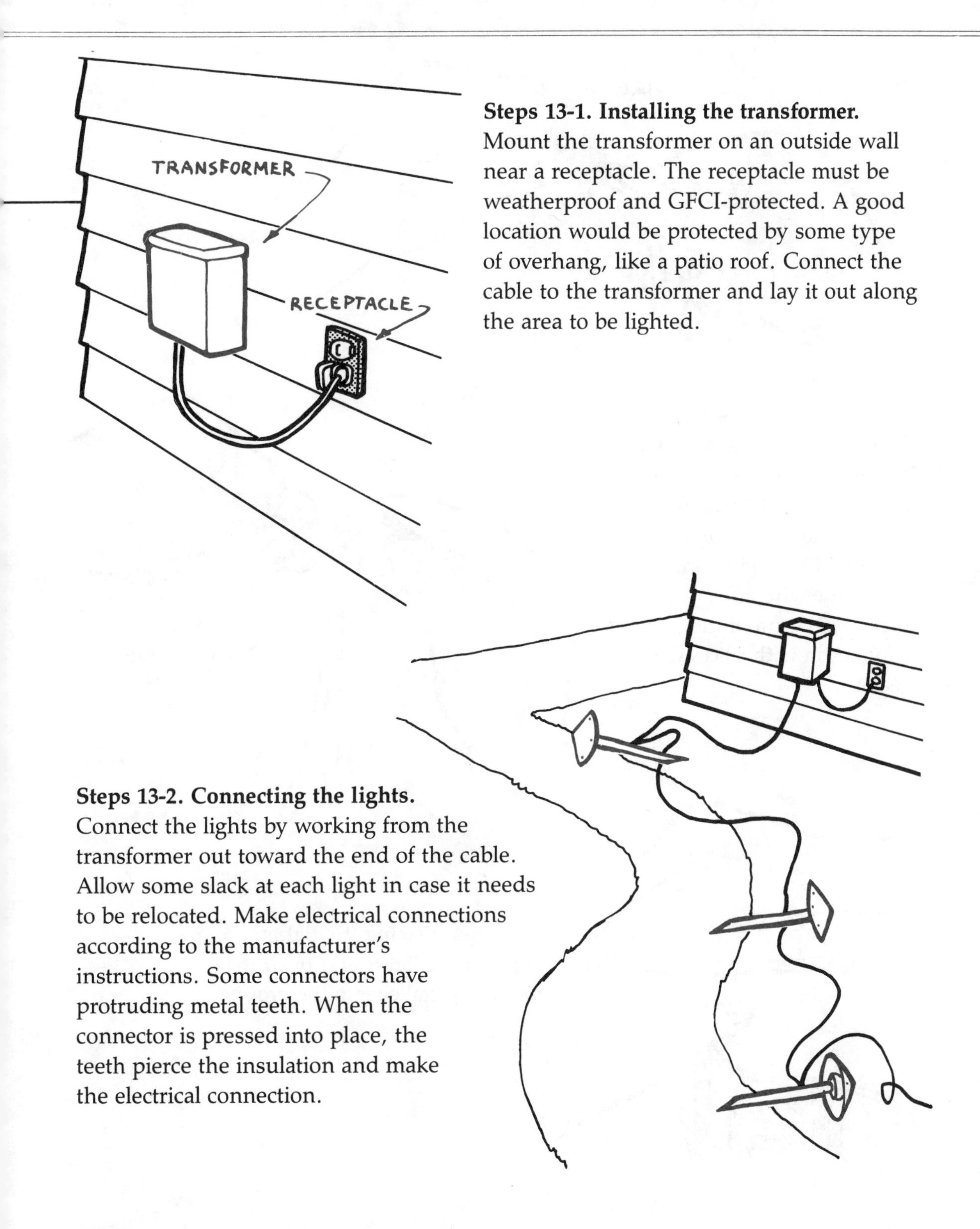


**Steps 13-1. Installing the transformer.** Mount the transformer on an outside wall near a receptacle. The receptacle must be weatherproof and GFCI-protected. A good location would be protected by some type of overhang, like a patio roof. Connect the cable to the transformer and lay it out along the area to be lighted.

**Steps 13-2. Connecting the lights.** Connect the lights by working from the transformer out toward the end of the cable. Allow some slack at each light in case it needs to be relocated. Make electrical connections according to the manufacturer's instructions. Some connectors have protruding metal teeth. When the connector is pressed into place, the teeth pierce the insulation and make the electrical connection.

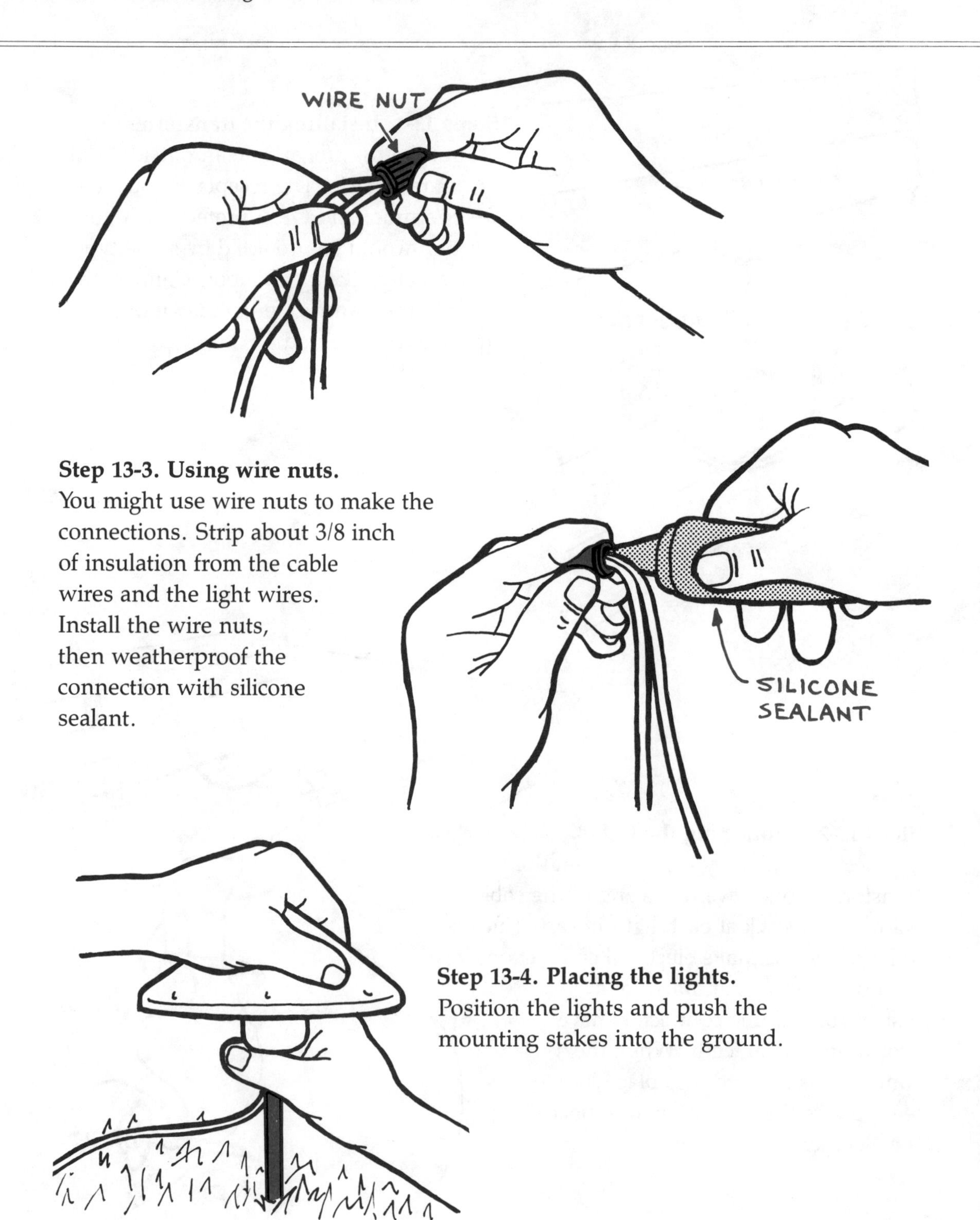


**Step 13-3. Using wire nuts.**
You might use wire nuts to make the connections. Strip about 3/8 inch of insulation from the cable wires and the light wires. Install the wire nuts, then weatherproof the connection with silicone sealant.

**Step 13-4. Placing the lights.**
Position the lights and push the mounting stakes into the ground.

**Step 13-5. Burying the cable.**
To bury the cable, press a flat spade several inches into the ground with one foot. Rock the spade back and forth until a small gap is opened. Continue working until you cut a narrow slit between the lights.

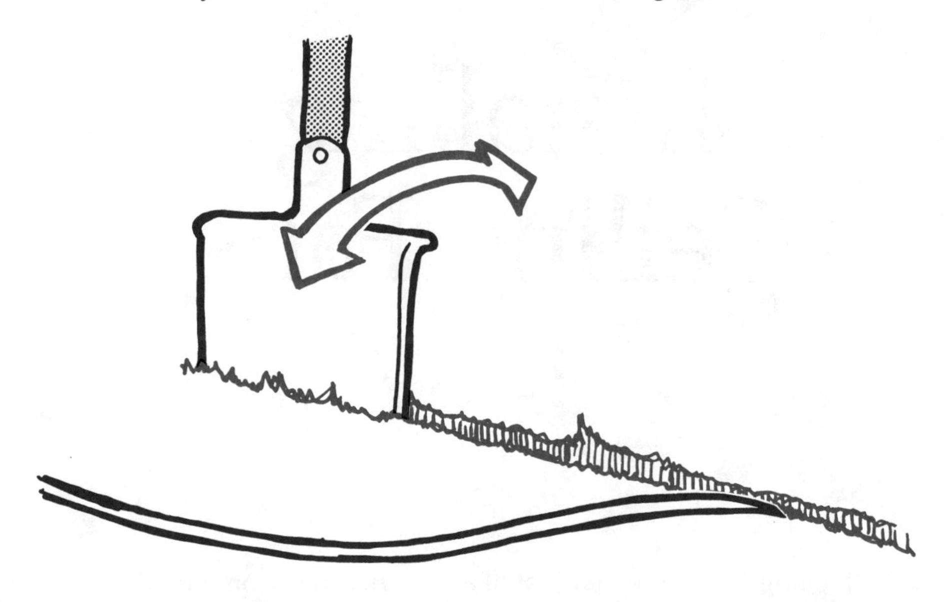

Place the cable into the slit and press the slit closed. Plug in the transformer and check the installation.

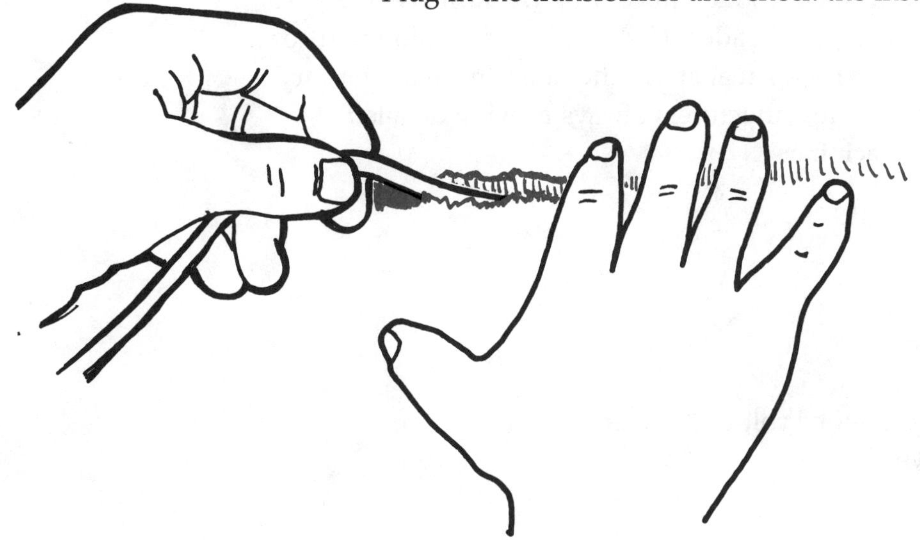

CHAPTER FOURTEEN

# Installing Ceiling Fans

Ceiling fans are wired just like light fixtures. Some models are very light and can be installed to existing outlet boxes, but heavier fans might need additional support. In this case, you should install a wooden header between the ceiling joists and mount the box on the header. The fan should be mounted so that the blades are at least 7 feet above the floor. Installations vary with each model, but manufacturers always provide detailed instructions with each fan.

## Tools & Materials

- ☐ Screwdriver
- ☐ Needle-nose pliers
- ☐ Wire stripper
- ☐ Wire nuts
- ☐ Neon voltage tester or volt-ohmmeter
- ☐ Adjustable wrench
- ☐ Ceiling fan

**Step 14-1. Types of fans.**
Some models have rods that extend the fan down from the ceiling; other models fit flush against the ceiling.

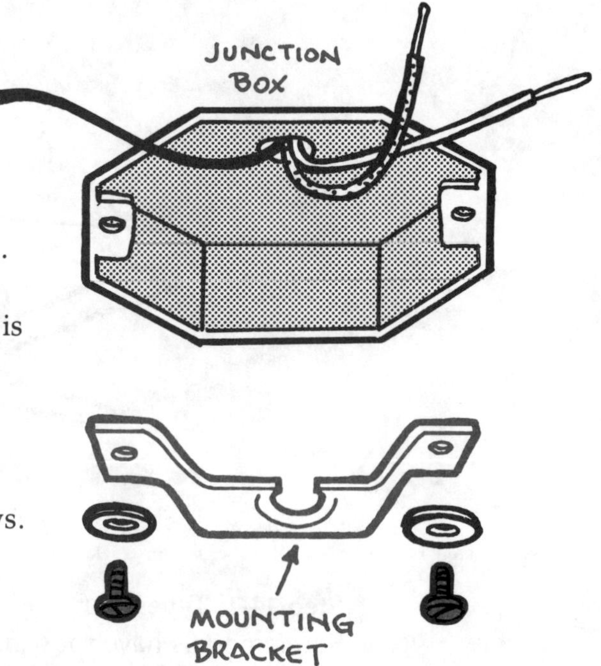


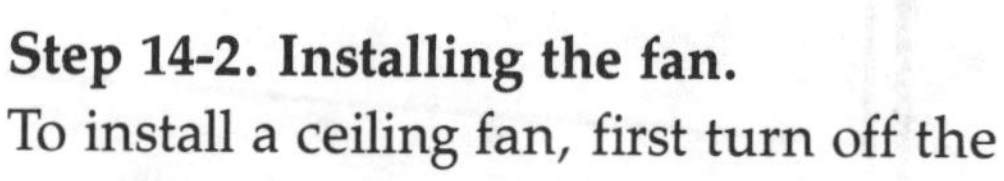

**Step 14-2. Installing the fan.**
To install a ceiling fan, first turn off the power to the circuit at the service panel. Before beginning work, use a voltage tester to make sure power to the circuit is off. Remove any existing fixture. Fasten the mounting bracket to a standard 4″×2 1/8″ metal octagon electrical box. Suspend the motor housing from the bracket and tighten the mounting screws.

**Step 14-3. Connecting wires.**
Use wire nuts to connect the house wires to the fan wires: black wire to black wire, white wire to white wire, and ground wire to green fan wire. As with any fixture, do not place any strain on the wires. Slide the canopy in place and install the canopy screws.

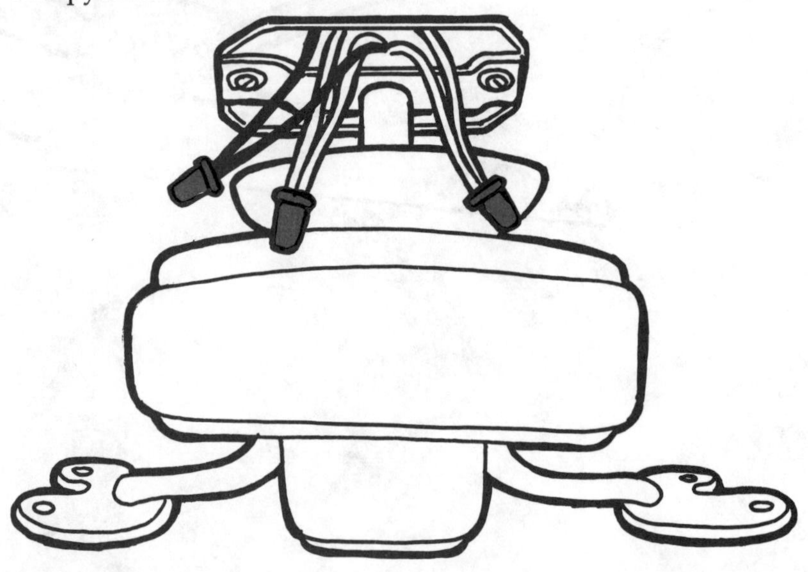

**Step 14-4. Attaching the blades.**
Fasten the blades to the blade holders using the provided screws, washers, and nuts. Restore the power and test the fan.

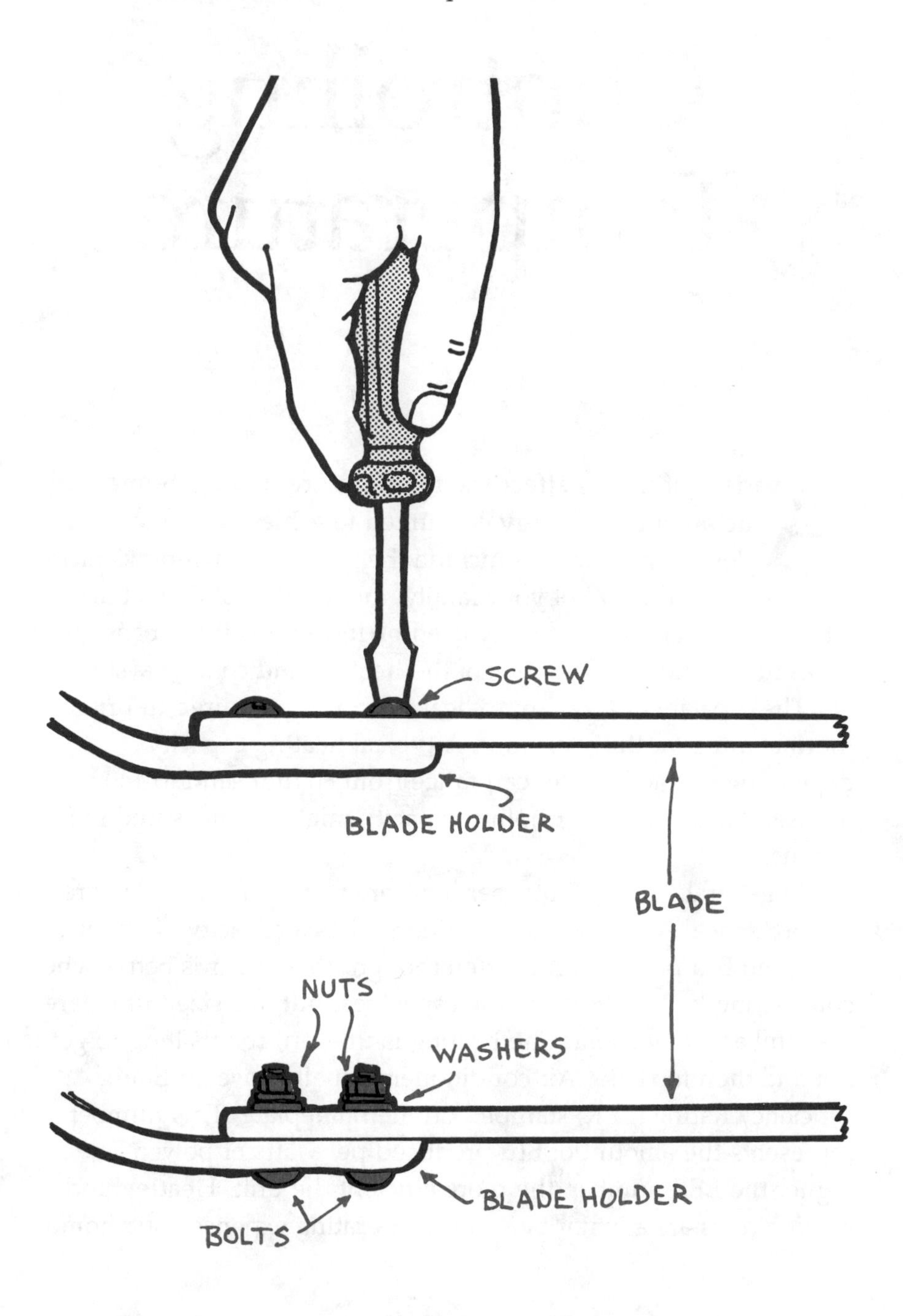

CHAPTER FIFTEEN

# Controlling Temperature

A variety of things affect the temperature in your home and the amount of energy consumed to achieve a comfortable level. These factors include the number of windows facing east and west, the size of your family, the number of lights burning (about 90 percent of the energy used by incandescent bulbs is given off as heat), and the efficiency of the heating and cooling system.

The capacity of a system to heat or cool is measured in British thermal units, or Btu, per hour. A typical heating capacity, depending on the climate, can range from 80 thousand to 140 thousand Btu per hour. Cooling capacity might be measured in Btu or tons.

In general, about 25 Btu per hour are needed for each square foot of floor space. A two-ton air conditioner has a capacity of about 24 thousand Btu per hour. It is often thought that larger is better when considering heating and cooling capacities, but oversized units are wasteful and inefficient. The heating is uneven; rooms tend to get too hot and then too cold. Air conditioners usually have an Energy Efficiency Rating (EER) stamped on the nameplate. This number represents the amount of Btu produced per watts of power. The higher the EER number, the more efficient the unit. Heating and cooling costs are a major part of the operating expenses of a home.

## Tools & Materials

- ☐ Screwdriver
- ☐ Adjustable wrench
- ☐ Needle-nose pliers
- ☐ Continuity tester
- ☐ Volt-ohmmeter
- ☐ Fin comb
- ☐ Vacuum cleaner

**Step 15-1. Thermostat coil.**
The heart of the comfort-control center is the *thermostat*. It operates on the principle that two different metals expand at different rates under the same temperature change. Two different metals are bonded together in the shape of a coiled spring.

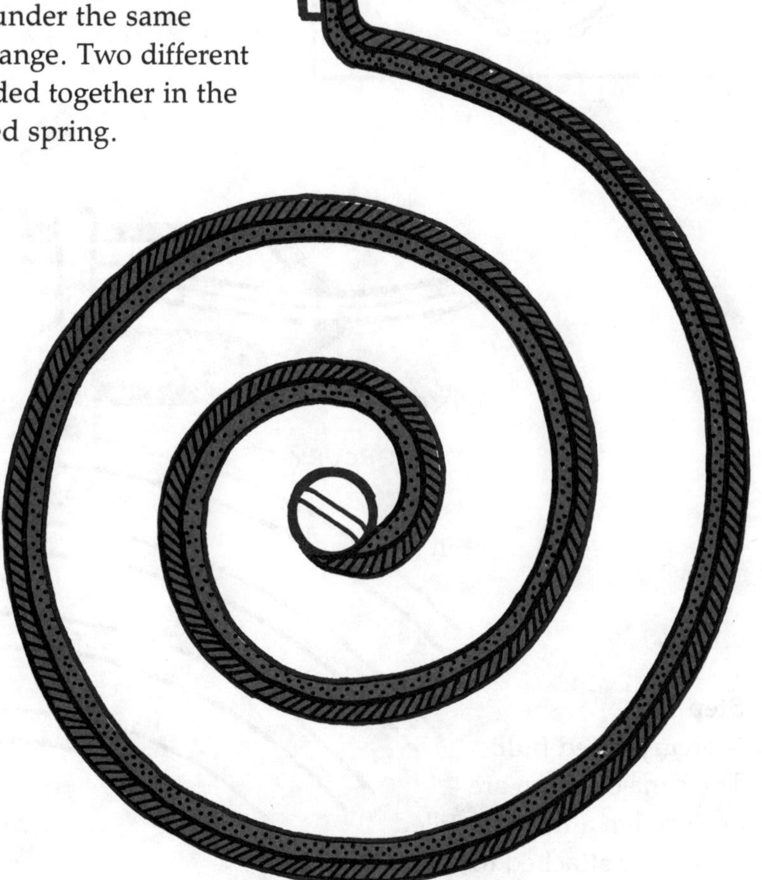

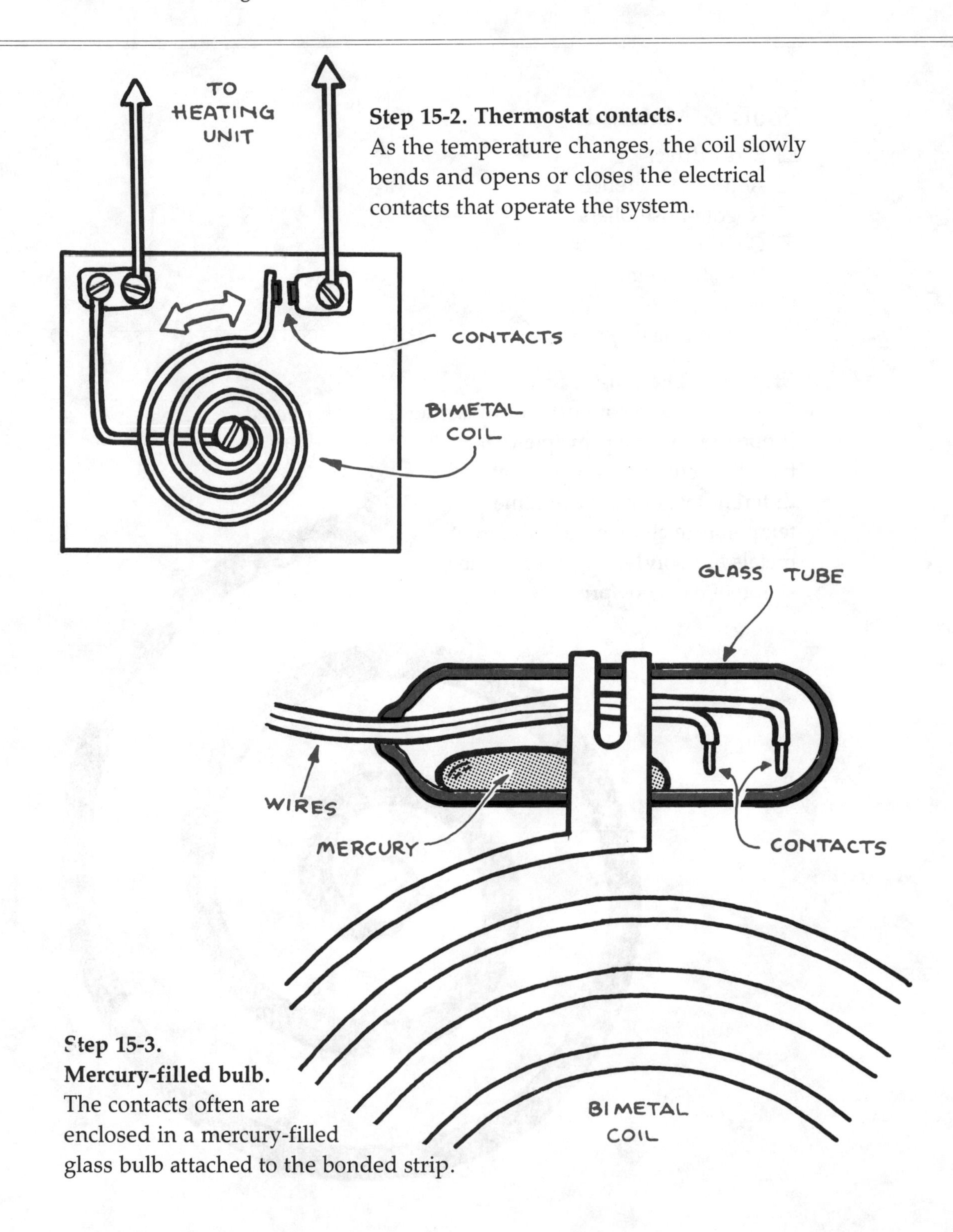


**Step 15-2. Thermostat contacts.**
As the temperature changes, the coil slowly bends and opens or closes the electrical contacts that operate the system.

**Step 15-3.**
**Mercury-filled bulb.**
The contacts often are enclosed in a mercury-filled glass bulb attached to the bonded strip.

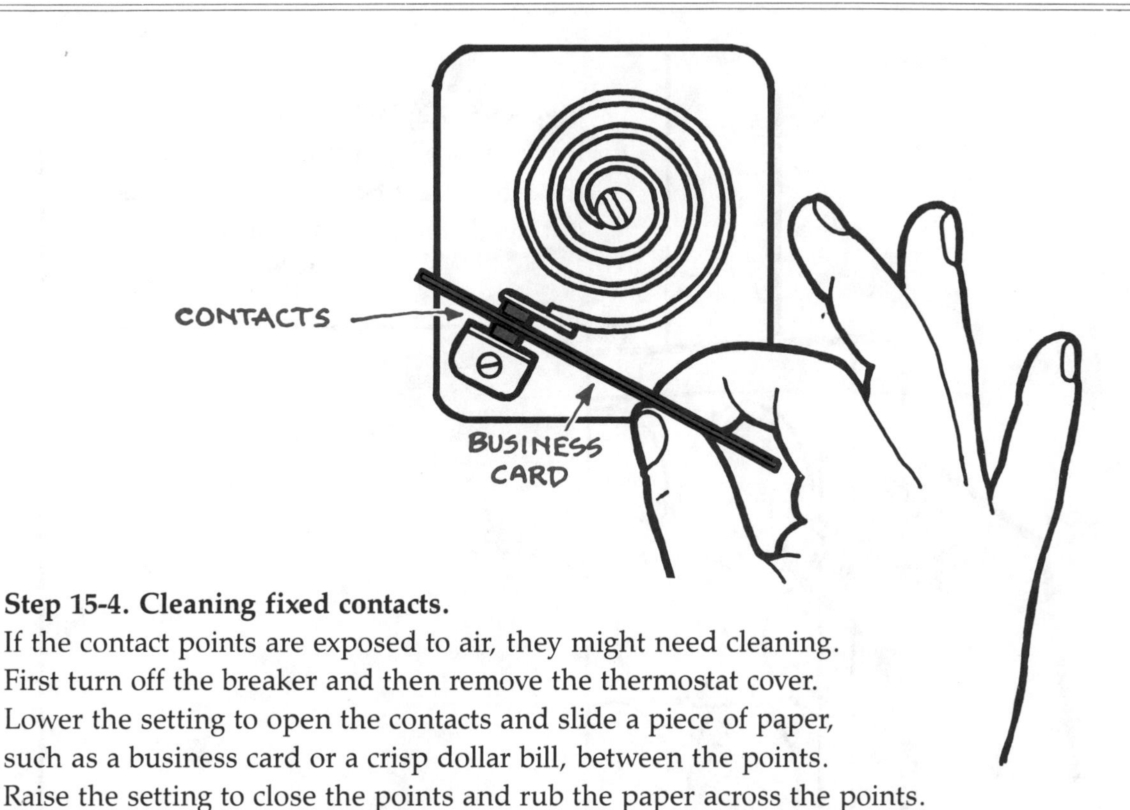


**Step 15-4. Cleaning fixed contacts.**
If the contact points are exposed to air, they might need cleaning. First turn off the breaker and then remove the thermostat cover. Lower the setting to open the contacts and slide a piece of paper, such as a business card or a crisp dollar bill, between the points. Raise the setting to close the points and rub the paper across the points.

**Step 15-5.**
**Cleaning variable contacts.**
Clean variable, or wiper, contacts by turning off the power and wiping the contacts with a cotton swab moistened in a solution of equal parts of water and vinegar. If the system fails to operate, check the breakers. Make sure it is receiving power. Then check the thermostat.

**Step 15-6. Replacing the filter.** If you notice insufficient air flow, the filter might be dirty. Check to see if it needs to be cleaned or replaced.

**Step 15-7. Tightening the belt.**
Insufficient air also can mean a loose blower drive belt.
Turn off the breaker and remove the panel to the blower.
Push down on the belt about half way between the pulleys.
It should give almost an inch if the tension is right.
If your system has a motor adjustment bolt, use a
wrench to turn the bolt until the tension is correct.

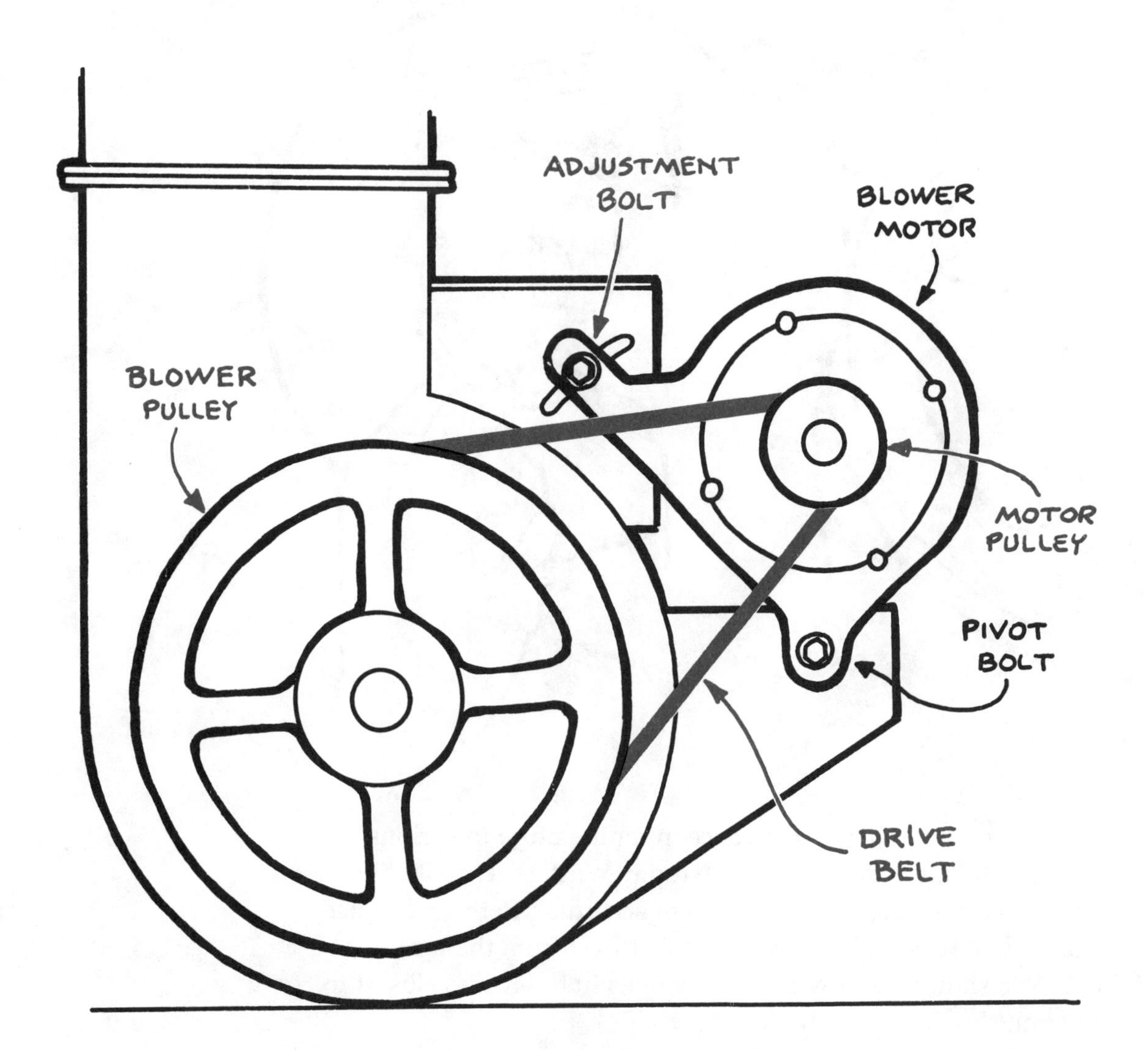

**Step 15-8. Replacing the belt.**
If your system doesn't have an adjustment bolt, and the belt is too loose, replace it. With the power still off, work the belt over the lip of the motor pulley. Keep one hand on the belt while turning the pulley with the other. Take the old belt with you when you buy the new one. Install the new belt in the reverse order.

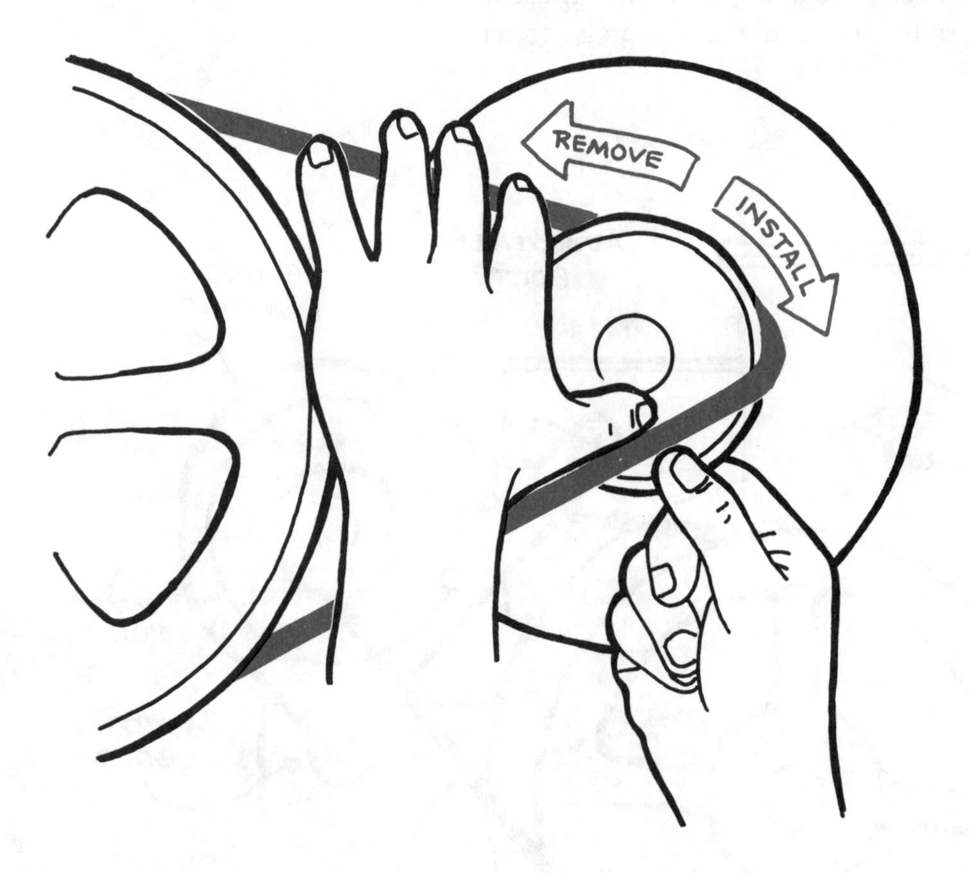


If you notice a large difference in temperature in various rooms, adjust the dampers in the room registers. Usually, the best way to set dampers is to adjust them according to their distance from the heating unit. In the rooms farthest from the unit, the dampers should be fully open. The ones in the rooms closest to the unit should be nearly closed.

**Step 15-9. Inside a window air conditioner.**
The heart of the cooling system is the compressor.

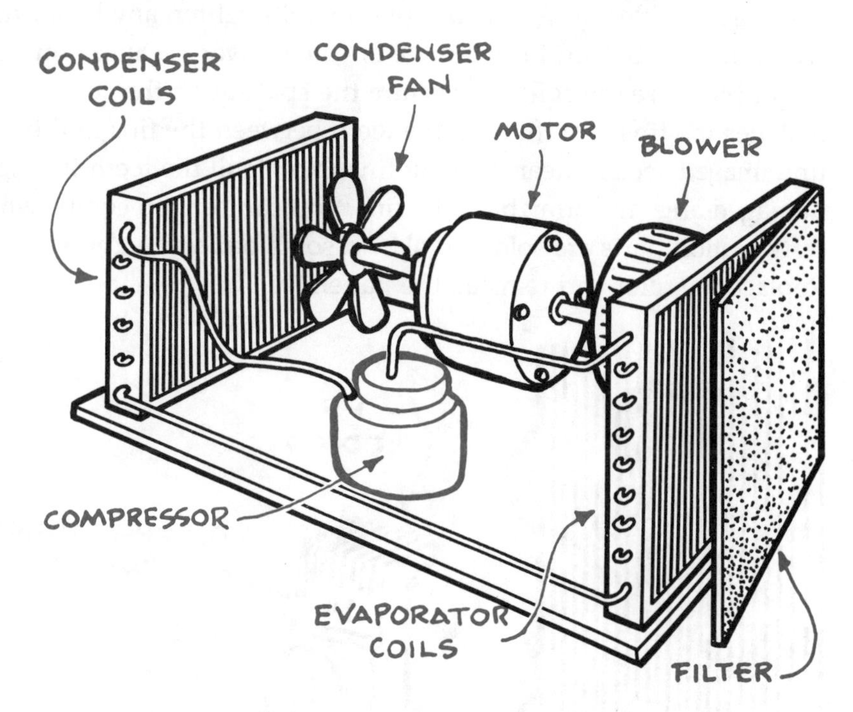


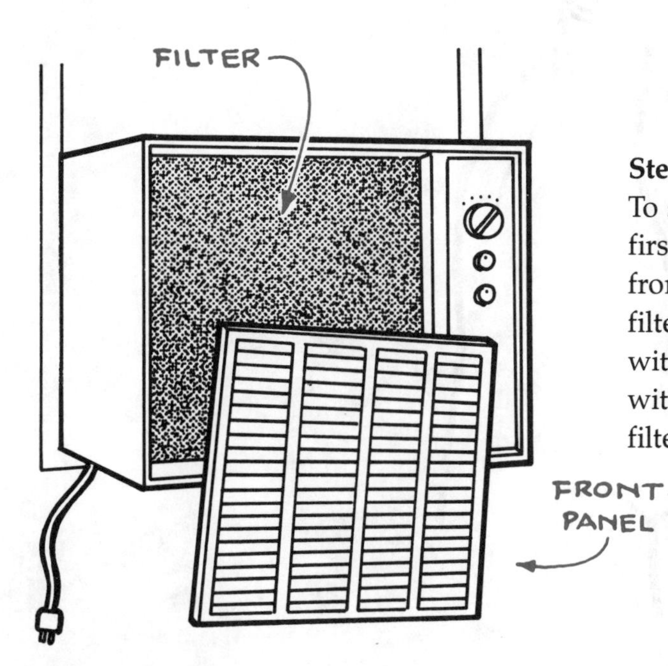


**Step 15-10. Cleaning a filter.**
To service a window air conditioner, first unplug the unit and remove the front cover. Remove the filter. Some filters can be cleaned by vacuuming with a brush attachment, then washing with mild soap and water. Rinse the filter thoroughly and shake it dry.

**Step 15-11. Straightening fins.**
Behind the filter is the evaporator. Use the brush attachment of your vacuum cleaner to vacuum the fins. Straighten any bent fins with a fin comb. Don't use a knife or screwdriver because you might puncture the coils. Make sure the spacing on the comb teeth matches the fins. Carefully fit the teeth between the fins in the undamaged section near the bent fins. Now pull the teeth through the damaged area and straighten the fins. You can buy fin combs where refrigeration parts are sold and also in some building supply stores. Replace the filter and reinstall the panel.

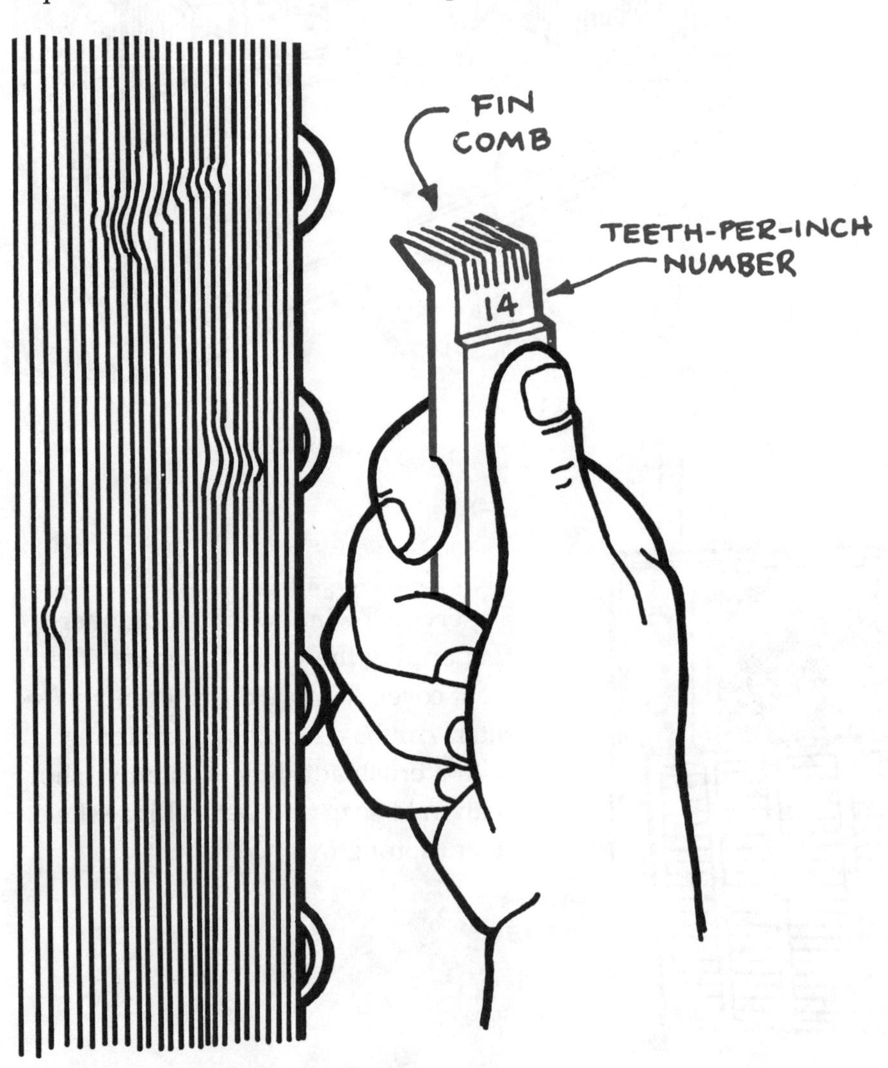

**Step 15-12. Checking the outside air vents.**
Use needle-nose pliers to remove any leaves or twigs. The fins behind the vents are cooling fins for the condenser coil. Be careful not to damage them. If the fins are exposed, straighten any bent ones with the fin comb. Plug the unit back in.

**Step 15-13. Checking the thermostat.**

If the compressor doesn't run, or runs continuously, the thermostat might be bad. To check it, unplug the air conditioner and remove any vent knobs. Open the control panel to gain access to the back of the thermostat. Unplug the two wires to the terminals on the thermostat. Connect a continuity tester to the terminals. Turn the control to the warmest setting. The tester light should not light. Now turn the control to the coldest setting. The tester light should light. If the thermostat fails this test, replace it.

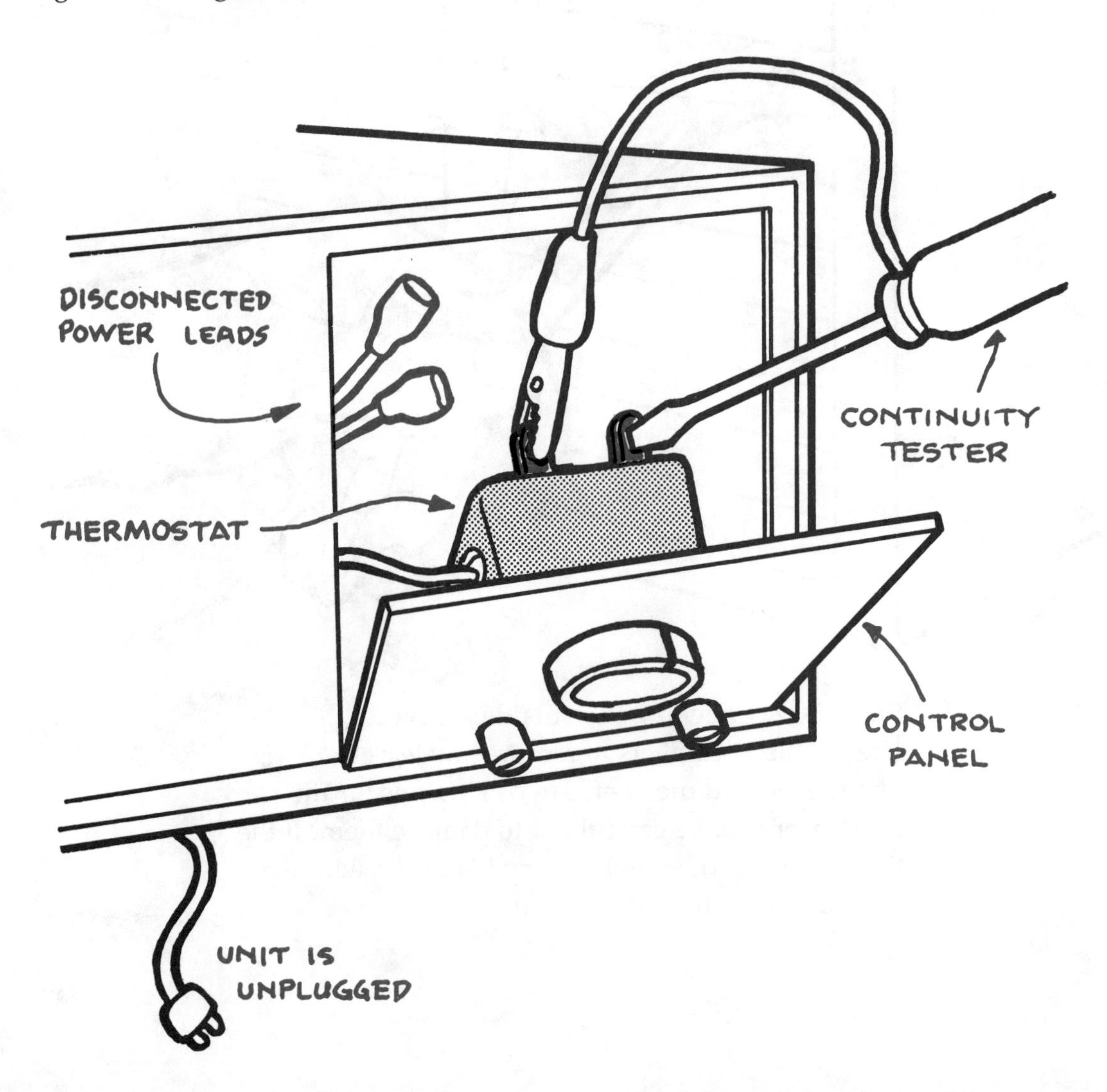

**Step 15-14. Replacing the thermostat.**
To replace the thermostat, pull off the control knob and remove the thermostat and sensing tubing. If possible, replace the thermostat with the same kind of unit. Be careful not to kink the new tube.

**Step 15-15. Central air conditioner.**
To service a central air conditioner, set the thermostat to the OFF position. Turn off the power to the unit at the disconnect switch and remove the top panel.

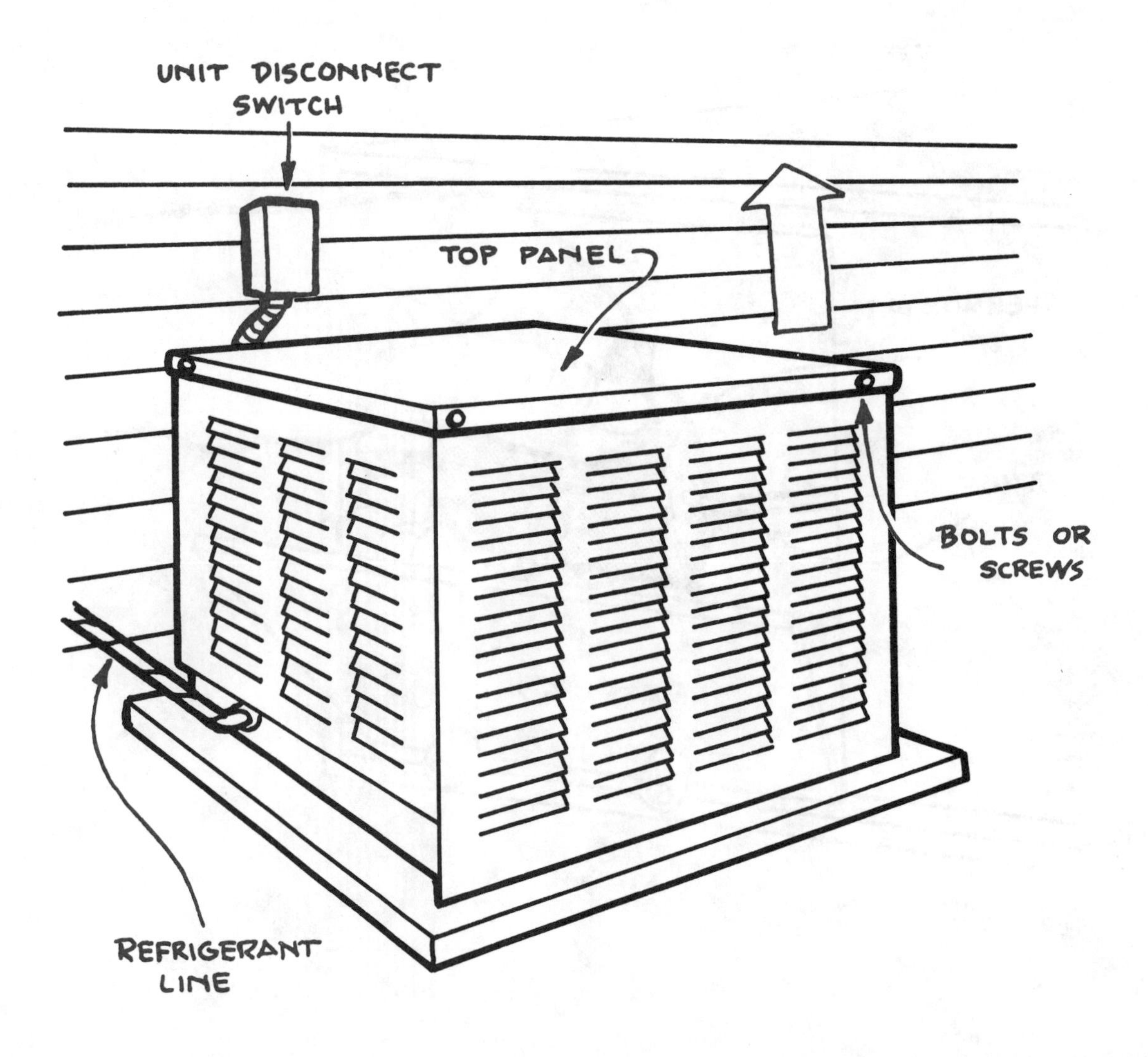

**Step 15-16. Straightening the fins.**
Use your vacuum's brush attachment to clean the coil fins. Straighten any bent fins with a fin comb.

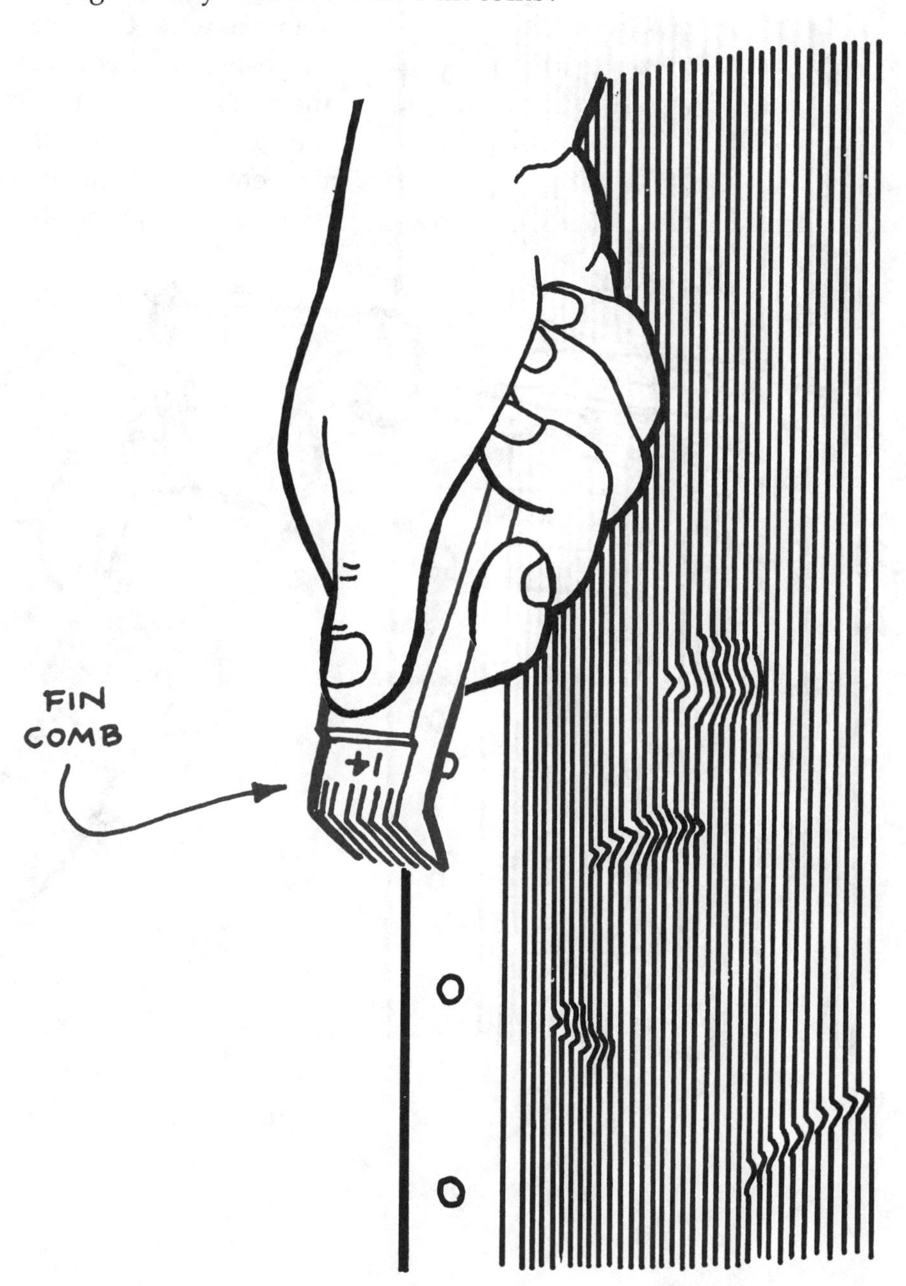

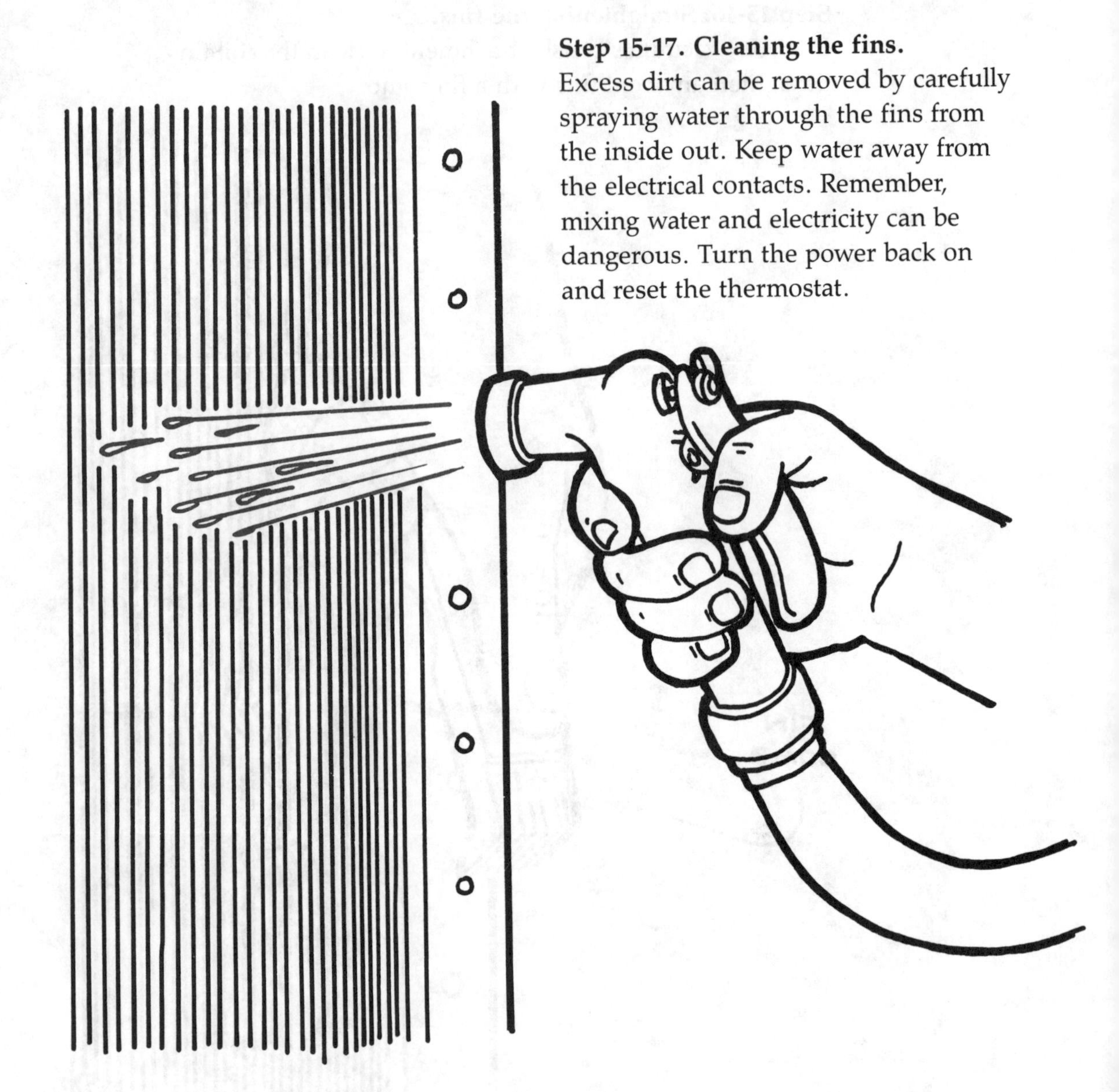

**Step 15-17. Cleaning the fins.**
Excess dirt can be removed by carefully spraying water through the fins from the inside out. Keep water away from the electrical contacts. Remember, mixing water and electricity can be dangerous. Turn the power back on and reset the thermostat.

# Glossary

**alternating current** Current that regularly reverses its direction, flowing first in one direction then the other. Abbreviated ac.

**ampere** A unit used in measuring electrical current. Abbreviated amp. It is based on the number of electrons flowing past a given point in one second. It can be determined by Ohm's law formula:

I (current in amps) = W (power in watts) ÷ V (voltage)

**AWG (American Wire Gauge)** The adopted standard of wire sizes, such as #12 wire, #14 wire, etc. The larger the number of the wire, the smaller the size of the wire. A #14 wire is smaller than a #12 wire.

**branch circuit** Any one of a number of separate circuits distributing electricity from an overcurrent protection device.

**cable** A stranded conductor or a group of individual conductors insulated from each other.

**cable, entrance** A heavy cable used to supply electrical service from the main line to a building.

**cable, nonmetallic sheathed** Two or more insulated wires assembled inside a plastic sheath. Type NM is used in dry locations. Type NMC might be used in both dry or damp locations.

**circuit** An electrical conductor forming a continuous path allowing current to flow from a power source through some device using electricity and back to the source.

**circuit breaker** A safety switch installed in a circuit that automatically interrupts the flow of electricity if the current exceeds a predetermined amount. Once tripped, a circuit breaker can be manually reset.

**conductor** The trade name for an electric wire or bus bar capable of carrying electricity.

**continuity** The state of having a continuous electrical path.

**current** The transfer of electrical energy caused by electrons traveling along a conductor. Abbreviated I, measured in amps.

**cycle** One complete reversal of alternating current where a forward flow (positive) is followed by a backward flow (negative). The standard rate in the United States is 60 cycles per second and is called 60 hertz (Hz).

**downstream/upstream** These terms refer to the location of a point in a circuit relative to the power source. Downstream refers to the part of the circuit from the point to the remaining part of the circuit going away from the power source. Upstream means the part of the circuit between the point and the power source.

**duplex receptacle** A receptacle providing electrical connections for two plugs.

**ground** A conducting connection between earth and an electrical circuit or equipment, whether intentional or accidental.

**hickey** A fitting used to support heavy, ceiling-mounted light fixtures such as chandeliers or fans.

**hot wire** The ungrounded current-carrying conductor of an electrical circuit. Normally, hot wires are identified by black or red insulation, but they can be any color except white, gray, or green.

**insulation** A nonconducting material used to cover wires and components to remove shock hazards and to prevent short circuits.

**low-voltage wiring** A method of wiring where a lower voltage is used to supply electricity for such purposes as door bells, thermostats, and some outdoor lighting.

**neutral bus bar** The solid metal bar in a service entrance panel or subpanel used as a common terminal to connect all of the neutral wires. The neutral bus bar in the service entrance panel is bonded to the panel as well as being directly connected with earth ground. The neutral bus bar in the subpanel is used to connect the neutral wires but is isolated from the panel and not grounded.

**neutral wire** The grounded conductor that provides the return path to the source completing the circuit. Neutral wires must never be interrupted by circuit breakers or fuses. Neutral wires are identified by white or gray insulation.

**ohm** The unit for measuring electrical resistance.

**ohmmeter** The instrument used in measuring electrical resistance.

**outlet** A point in a wiring system where current can be taken to supply electrical equipment.

**receptacle** A connecting device designed to accept plugs.

**resistance** The property in an electrical conductor or circuit that restricts the flow of current. It is measured in units called ohms.

**service entrance panel** The main power cabinet containing the main breaker and circuit breakers distributing electricity throughout the residence.

**short circuit** A completed, very low-resistance circuit, where two bare hot wires come in contact or a bare hot wire touches a bare ground wire or grounded component.

**switch** A device used to close or open a circuit allowing current to flow or not flow.

**terminal** A point used to make electrical connections.

**thermostat** A device used to control temperatures to a predetermined level.

**transformer** A device used to transfer electrical energy from one circuit to another using electromagnetic induction. Two types commonly found are the step-up and step-down transformers. The step-up transformer is used where a higher output voltage than the input voltage is desired. A step-down transformer is used for circuits where a lower output voltage than the input voltage is needed, as with doorbells and thermostats.

**volt** Unit used in measuring electrical pressure. Abbreviated V.

**voltage** The electrical pressure, measured in volts, at which a circuit operates. Voltage can be determined by the Ohm's law formula:

W (power in watts) ÷ I (current in amps) = V (voltage)

**voltmeter** An instrument used to measure voltage.

**volt-ohmmeter** An instrument that measures both voltage and electrical resistance.

**watt** The unit used in measuring electrical power. Abbreviated W. The amount of power in a circuit can be determined by the Ohm's law formula:

I (current in amps) × V (voltage) = W (power in watts)

# Index

## D



## E



## F



## G



## H



## I



## J



## K

## L



## M



## N



## O



## P



## Q



## R

# Shopping List for All Thumbs Home Wiring

- ❑ Volt-ohmmeter
- ❑ Neon voltage tester
- ❑ Continuity tester
- ❑ Diagonal wire cutters (dikes)
- ❑ Needle-nose pliers
- ❑ Standard flat-blade screwdriver 3/8″
- ❑ Medium Phillips screwdriver
- ❑ Small Phillips screwdriver
- ❑ Wire stripper
- ❑ Cable ripper
- ❑ Electrical tape
- ❑ Package of wire nuts for #14 and #12 wire
- ❑ Adjustable wrench
- ❑ Crimping tool
- ❑ Fin comb
- ❑ Hammer
- ❑ ____________________
- ❑ ____________________
- ❑ ____________________
- ❑ ____________________

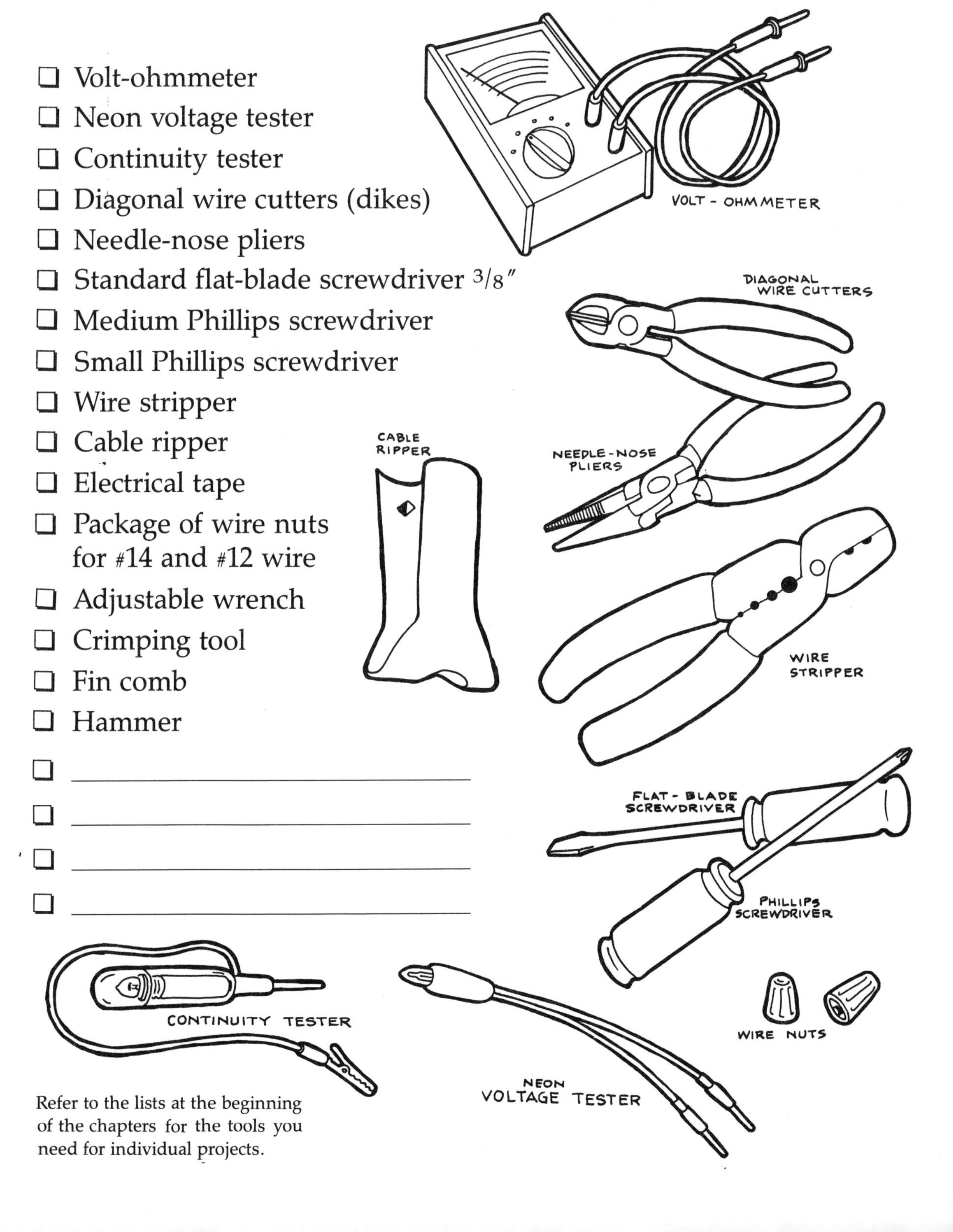


Refer to the lists at the beginning of the chapters for the tools you need for individual projects.

# Safety Tips

Before you start:

- Think safety.
- Work slowly and carefully.
- Read and understand all instructions.
- Gather all your tools and required materials.
- Make sure your test equipment is working properly.
- Familiarize yourself with how your home is wired. Identify and label circuits in the service panel for future reference.
- Never work on any live circuit, fixture, or appliance.
- Turn off the power or unplug the appliance.
- Use your test equipment to confirm that the power is off.
- If you have any doubt that the power to a circuit is off, turn off the main breaker.
- Make sure that no one will restore the power while you are working. One way to do this is to tag the breaker with a notice reading "Danger! Do not reset."
- Make sure you know the procedures to follow if a person comes in contact with a live circuit. If someone has been shocked and is still part of the live circuit, don't touch him or her with your bare hands. First try to disconnect the power. If you cannot, use some type of insulated device, such as a coat or broom, to remove the victim from the circuit. Keep the victim warm while you call for help. You might need to perform artificial respiration.